Wind

ENERGY BASICS

A Guide to Small and Micro Wind Systems

PAUL GIPE

CHELSEA GREEN PUBLISHING COMPANY
WHITE RIVER JUNCTION, VERMONT

621.31
G514w
1999

Designed by Kate Mueller.
Cover designed by Ann Aspell.

Printed in the United States.
First printing, April 1999.

06 05 04 03   5 6 7 8

Disclaimer
The installation and operation of small wind turbines entails a degree of risk. Always consult the manufacturer and applicable building codes before installing or operating your wind power system. When planning or installing wind turbines that will be interconnected with an electric utility, always check with the utility first. When in doubt, ask for advice. Suggestions in this book are not a substitute for common sense or the instructions from wind turbine manufacturers and regulatory agencies. The author assumes no liability for personal injury, property damage, or loss arising from information contained in this book.

Library of Congress Cataloging-in-Publication Data
    Gipe, Paul
    Wind energy basics: a guide to small and micro wind systems / Paul Gipe.
       p.   cm.   —(Real Goods solar living book)
    Includes bibliographical references and index.
    ISBN 1-890132-07-1 (alk. paper)
    1. Wind power.   I. Title.   II. Series.
TJ820.G55   1999
621.31'2136—dc21   98-54585

Chelsea Green Publishing Company
Post Office Box 428
White River Junction, VT 05001
(800) 639-4099
www.chelseagreen.com

Wind
ENERGY BASICS

Dedication

For Ed Wulf, a man who saw the future.

Contents

FOREWORD   ix

PREFACE   xi

OVERVIEW   1

1. FUNDAMENTALS   7
   Power in the Wind   7
   Swept Area   9
   Wind Speed Distribution   10
   Wind Resources   11
   Increases in Wind Speed and
      Power with Height   12
      Units of Measurement   12
      Relative Size   13

2. ESTIMATING PERFORMANCE   17
   Swept Area Method   18
   Power Curve Method   20
   Manufacturer's Estimates   21

3. WIND TURBINE TECHNOLOGY   25
   Configurations   25
      Two or Three Blades   25
      Blade Materials   25
      Orientation   26
      Robustness   26

   Overspeed Control   26
   Generators   28
   Micro Turbines   29
      LVM Aerogen   31
      Marlec 913   31
      Ampair 100   32
      Southwest Windpower Air 303   32
      Aerocraft   34
   Mini Turbines   34
      World Power Technologies   35
      Bergey Windpower   36
   Household-size Turbines   36
      Proven Engineering   38
      Wind Turbine Industries   38
   Novel Turbines   38
   Towers   40
      Rooftop Mounting   41

4. OFF-THE-GRID APPLICATIONS   43
   Hybrid Wind and Solar Systems   44
   Wind Pumping   48
   Wind Heating   50
   Recreational Vehicles   54
   Cabins and Cottages   54
   Electric Fence Charging   57
   Electric Vehicle Charging   57
   Telecommunications   57
   Village Power   59

5. INTERCONNECTED OR UTILITY INTERTIE   61
Interconnection Technology   63
   Induction Generators   63
   Inverters   64
Degree of Self-use   65
Net Metering   67
Power Quality and the Utility   67
European Distributed Generation   67

6. SITING AND SAFETY   71
Tower Height   71
Tower Placement   75
Noise   75
Urban Wind   75
Safety   76
   Moving Machinery   76
   Electrical   77
   Batteries   77
   Towers   77
   Servicing   78

7. BUYING SMALL TURBINES   81
Towers   82
Control Panels   83
Price   84
Do-it-yourself Turbines   85
Used Wind Turbines   85
Collective Buying Power   86
Subsidies   87

8. INSTALLING SMALL TURBINES   89
Tower Raising   89
Griphoists   90
Choosing a Griphoist   93
Tower Conductors   93

Maintenance   95
Raising a Small Turbine   95

9. CONCLUSION   103

APPENDIX 1: CHARACTERISTICS OF SELECTED SMALL TURBINES   106
Table A: Selected Micro Turbines   106
Table B: Selected Mini Turbines   106
Table C: Selected Household-size Turbines   108

APPENDIX 2: RESOURCES   110
Selected Small Wind Turbine Manufacturers   110
Guyed Towers for Micro and Mini Wind Turbines   111
Reconditioned Small Turbines   112
Selected Manufacturers and Distributors of Wind Pumps   112
Wind Measurement Devices   112
Inverters and Controls   113
Griphoists   113
Wind Energy Associations   113
Sources of Wind Data   114
Mail-order Catalogs   115
Publications   115
   Plans   115
   Books   116
   Magazines   117
   Market Surveys   117
   Videos   117
Electronic Information Sources   117

INDEX   118

Foreword

With the use of renewable energy (RE) resources on the rise, the number of wind power systems in use is exploding. There are currently over 150,000 small-scale RE systems in America, and this number grows by 30 percent annually. The small-scale use of wind power is growing at twice that rate—over 60 percent per year. What Americans and folks all over the world are finding out is that wind power is an excellent and cost-effective alternative to utility line extensions, utility bills, and fossil-fuel generators.

Many of these small wind turbines are finding jobs in hybrid systems, making power right alongside their RE cousins, the photo-voltaics (PV). According to our research at *Home Power*, 82 percent of the small wind turbines work in systems that also contain PVs. This is a marriage made in Heaven: In most locations, when the sun doesn't shine the wind most certainly blows.

Small wind turbines allow RE users to further reduce their need for engine generators or utility power, making them more self-sufficient, reliable, and less polluting. In *Home Power*'s specific case, our 1.8 kW PV array produces power right alongside our Whisper H1500 wind turbine. This marriage of RE resources has reduced our winter engine generator operating time by over 70 percent.

With literally three dozen different small wind generators to choose from, this book fills the information void. Within these pages you will find the information you need to put the right wind turbine to work at your homestead.

May the sun shine, the wind blow, and the water run!

Richard and Karen Perez
Home Power Magazine
Written on 13 May 1998 at
Funky Mountain Institute
using only solar and wind power.

Preface

ind Energy Basics is a small book about small wind turbines. It is not by any means exhaustive, nor is it intended to be. In the more than two decades I've worked with wind energy, the field has grown so vast that it's no longer possible to confine the technology within the covers of one book, even after limiting it only to small wind turbines.

Wind Energy Basics is intended to be a companion to *Wind Power for Home & Business* (Chelsea Green, 1993) and *Wind Energy Comes of Age* (John Wiley & Sons, 1995). *Wind Energy Basics* introduces micro and mini wind turbines and describes how to use them. *Wind Power for Home & Business*, this book's big brother, provides a comprehensive discourse on small wind turbines and introduces medium-size wind turbines. *Wind Energy Comes of Age* examines the commercial success of medium-size wind turbines and the phenomenal growth of the wind energy industry worldwide.

While small wind turbines have yet to reach the status of technological commodities such as personal computers, the debut of micro and mini wind turbines brings the technology within reach of almost everyone. These inexpensive machines, when coupled with readily available photovoltaic panels, have revolutionized living off-the-grid. And their increasing popularity has opened up new applications previously considered off-limits to small wind turbines, such as charging electric fences in Denmark and powering remote telephone call boxes in Great Britain.

All books, even small ones, require the help and cooperation of many people. My thanks to the many small wind turbine manufacturers worldwide who answered my frequent queries about their products; to Mick Sagrillo, Hugh Piggott, and Eric Eggleston for their comments and insights into small wind turbine design; and to Nancy Nies for her help installing our own mini wind turbine. And my heartfelt gratitude to the Folkecenter for Renewable Energy and the people of Denmark for a fellowship to study the distributed use of wind energy in northwest Jutland.

God vind! (Good Wind!)
Paul Gipe
Tehachapi, California
January 1999

Small wind turbines undergoing tests at the Folkecenter for Renewable Energy in Denmark.

Overview

Wind energy technology is booming. Not since the heyday of the farm windmill has the use of wind energy grown at such a dramatic pace. By the new millennium, more than 40,000 medium-size wind turbines will be in operation worldwide, mostly in California, Europe, and India. These commercial wind turbines, including those found in California's giant wind power plants, will produce 20 terawatt-hours (20 billion kilowatt-hours) of wind-generated electricity annually—enough to meet the needs of more than three million Californians, or twice as many Europeans, who are typically more energy-conscious.

But the success of medium-size wind turbines is only part of the story. Small wind turbines have found their role expanding as well. Whether it's on the contemporary homestead of Ed Wulf in California's Tehachapi Mountains, in the Chilean village of Puaucho overlooking the Pacific Ocean, or on the Scoraig peninsula of Scotland's wind-swept west coast, small wind turbines are making an important difference. While their contributions may be small in absolute terms, small wind turbines make a big difference in the daily lives of people in remote areas around the globe.

Small wind turbines, like those discussed in this book, may produce only a small number of kilowatt-hours (kWh) per month, but this electricity goes much further and provides as much, if not more, value to those who depend upon it as the generation of their bigger brethren.

> While their contributions may be small in absolute terms, small wind turbines make a big difference in the daily lives of people in remote areas around the globe.

Today there are more than fifty manufacturers of small wind turbines worldwide, and they produce more than one hundred different models. Altogether manufacturers in western countries have built about sixty thousand small wind turbines during the last two

Micro wind turbine. *Marlec Rutland 910 powers a yurt on the Mongolian steppes. Micro wind turbines have revolutionized the use of wind energy, bringing it within reach of almost everyone (Peter Fraenkel).*

Mini wind turbine. *Whisper 600 overlooking a winter scene on Lake Superior in Duluth, Minnesota. World Power Technologies' Whisper line of battery-charging mini wind turbines can be found powering cabins and vacation homes in many rural areas of North America.*

decades. And tens of thousands more have been manufactured in China for use by nomads on the Mongolian steppes.

The largest number of wind power machines installed fall into the micro wind turbine category. These are wind turbines so small you can carry them in your hands. Though micro turbines have been around for decades for use on sailboats, they have only gained prominence in the 1990s as their broader potential for off-the-grid applications on land has become more widely known.

Southwest Windpower, a wind turbine manufacturer, awakened latent consumer interest in micro turbines with the introduction of its sleek Air 303. Since its debut in 1995, Southwest Windpower has shipped eighteen thousand of the popular and inexpensive turbines. Southwest's Air 303 has quickly gained ground on the stalwarts of micro turbine manufacturers Marlec, LVM, and Ampair. Marlec has built more than twenty thousand of its multiblade Rutland model during the 1980s and early 1990s, mostly for the marine market. Not far behind is LVM, who has built some fifteen thousand multiblade turbines. Ampair, another British manufacturer of micro turbines for boaters, has shipped more than five thousand of their signature product, the Ampair 100.

The number of the more familiar household-size turbines built by companies such as Bergey Windpower and World Power Technologies, though significant, is far smaller. Between them, the two American producers have built more than four thousand small wind turbines not unlike the windchargers once used throughout the Great Plains.

During the 1930s several midwestern manufacturers built small battery-charging wind turbines for remote homesteads on the

American steppes that stretch from Texas to Alberta. Often, these windchargers were the only source of electricity for many farms and ranches in the days before rural electrification. Thousands of wind turbines, producing a few hundred watts to several kilowatts, were built. Familiar names like Sears, Zenith, and Montgomery Ward could be found emblazoned on their tail vanes, along with Wincharger, Jacobs, and the less well-known Parris-Dunn.

With the advent of rural electrification, the windcharger industry collapsed in the late 1940s and early 1950s. It wasn't until the 1970s that the oil crisis spurred renewed interest in small wind turbines. In those

Rebuilt Jacobs. *Alternative Energy Institute's Ken Starcher (left) and the author (right) replace the brushes on a reconditioned Jacobs windcharger. The highly valued Jacobs windcharger spurred a revival of small wind turbine use in the late 1970s. Many of the rebuilt units are still operating today.*

days the shortest route to wind energy was salvaging a 1930s-era windcharger and rebuilding it. Many did, and rebuilt Jacobs and Wincharger turbines, some now fifty years old, are still in service. It's not uncommon even today to find advertisements for these machines in the classifieds of *Home Power* magazine.

During the 1980s, manufacturers developed new designs for small wind turbines, incorporating lessons learned from the windcharger period. Most of the new battery-charging turbines switched from direct current (DC) generators to permanent-magnet alternators. The alternating current (AC) of these machines is rectified to DC for charging batteries or powering an inverter.

Several manufacturers in the 1980s built small wind turbines using induction genera-

tors for direct interconnection with an electric utility's power lines. Though technically an elegant solution for integrating wind turbines with the utility grid, small interconnected wind turbines proved a commercial failure in the United States for regulatory and political reasons.

In Europe, however, small wind turbines driving induction generators—known to electrical engineers as asynchronous generators—found a much more receptive climate, especially in Denmark, Germany, and the Netherlands. Experimenters, hobbyists, and small metalworking shops began building small wind turbines in the mid-1970s. Designed to supplement electricity from the local utility, the wind turbines gradually grew in size, from 10 kilowatts (kW), to 15 kW, then 30 kW. By 1982, they reached the then

Medium-size wind turbines. *Wind turbines from 15-kilowatt to 600-kilowatt power homes and farms throughout Denmark and Germany. Excess electricity is sold to the local utility. Of the thousands of wind machines operating in Denmark, there are several hundred household-size turbines like the locally built Thy Mølle in the foreground.*

incredible size of 50 kW. This grassroots effort spawned a billion dollar industry that today builds wind turbines from 500 kW to 1.5 megawatts.

Many of the wind turbines in northern Europe stand individually or in small clusters on the farms where their electricity is used. Today each one of these modern wind turbines is capable of meeting the residential needs of hundreds of homes. Though these machines long ago crossed the threshold from small to medium-size wind turbines, their use and popularity for providing home wind power is an important part of the wind energy story.

Household-size wind turbine. *Small wind turbines, such as this Bergey Excel in Tehachapi, Caifornia, can meet the needs of entire households.*

Wincharger Giant. *1930s-era windcharger standing sentinel on Mormon Row, Grand Teton National Park. One of the thousands of windchargers used on the American Great Plains prior to rural electrification.*

1
Fundamentals

There is just no escaping the fact that the amount of wind you have at your site determines how much power you can expect from a wind turbine of a given size. Though few people would ever consider placing a solar panel in the shade and expecting it to work, it's surprising when people try the equivalent with their wind turbine.

POWER IN THE WIND

The power (P) in the wind is a function of air density (ρ), the area intercepting the wind (A), and the instantaneous wind velocity (V), or speed. Increase any one of these factors and you increase the power available from the wind.

$$Power = \tfrac{1}{2}\rho AV^3$$

If the value for air density at sea level is substituted for ρ in the equation, power in watts is

P = 0.6125 AV^3, where speed is in meters per second (m/s) and area is in square meters (m^2), or

P = 0.00508 AV^3, where wind speed is in miles per hour (mph) and area is in square feet (ft^2).

Air density varies with temperature and elevation. Warm air is less dense than cold air. Any given wind turbine will produce less in the heat of summer than it will in the dead of winter with winds of the same speed. Minnesotans seeking wind development to offset an aging nuclear plant proudly boast that the upper Midwest's frigid winter winds hold more power than the equivalent winds in hot southern California. Similarly, there is less power in the wind for a specific wind speed at a mountaintop telecommunications site in Idaho than near sea level at the Folkecenter for Renewable Energy on Skibsted Fjord in Denmark.

Changes in air density relative to standard conditions at sea level can cut power production 10 to 20 percent, sometimes even more. For example, on a summer day atop 5,000-foot (1,500-meter) Cameron Ridge in the Tehachapi Pass the air temperature can reach 95°F (35°C). Under these conditions, the air

Anemometer mast. *Easily erected guyed anemometer masts of thin-walled tubing can be used to measure wind speed, or better yet, can be used to support a micro turbine: bird's-eye view (NRG Systems).*

is 80 percent as dense as along the coast at a comfortable 68°F (20°C). But the effect of changes in temperature or elevation on wind power, especially on small wind turbines, is dwarfed by changes in wind speed.

Power in the wind varies with the cube of wind speed. Double the speed of the wind and you increase the power available by eight times. Even a small increase in wind speed substantially boosts the power in the wind. This is the reason for the emphasis on putting your turbine where the winds are best.

Consider the power available at one site with a wind speed of 10 (the units don't matter here) and another site with a wind speed of 12. The wind at the windier site is only 20 percent greater ($^{12}/_{10}$ = 1.2), yet there is 70 percent more power available.

$$P_2/P_1 = (V_2/V_1)^3$$

$$P_2 = (^{12}/_{10})^3 P_1 = 1.73\ P_1$$

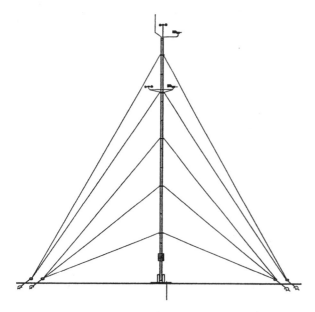

Sketch of guyed mast (NRG Systems).

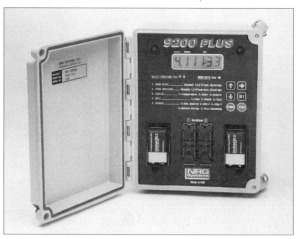

Wind data logger. *Electronic data recorders can be used to collect wind data or, with the appropriate sensors, monitor small wind turbine performance (NRG Systems).*

Because small changes in wind speed have such a profound effect on the power in the wind, many in the past have recommended measuring wind speeds for a minimum of one year prior to deciding whether or not to install a wind turbine. Commercial wind power developers in many parts of the world do just that. They install an anemometer and monitor the winds with an electronic data logger.

This is impractical for most people who might use a small wind turbine, simply because the cost of the mast, measuring devices, and professional analysis needed will often exceed the cost of a small wind turbine. For years Mick Sagrillo, a small wind turbine guru in Wisconsin, has suggested that it is better to install a micro turbine and monitor its performance. If the owner is dissatisfied with its production, it can be removed and sold. There is more of a market for used wind turbines than for used anemometers. Another approach is to locate owners of small wind turbines in your vicinity and determine how well their machines have performed.

Wind speeds vary over time. Energy, which is what we're really after, is the product of power and time. We'll come back to wind speeds shortly.

SWEPT AREA

As we've seen, the power in the wind is exponentially related to wind speed. Power is also directly related to the area intercepting the wind, that is, the area swept by the wind turbine's rotor. Double this area and you double the power available. Consider a conventional wind turbine, where the rotor spins about a horizontal axis. The rotor sweeps a disc the area of a circle,

$$A = \pi R^2$$

where area (A) equals the product of π and the square of the rotor's radius (R), or about the length of one blade. This formula gives the area (A) of the windstream swept or intercepted by the rotor.

This relationship between the rotor's radius (or diameter) and energy capture is fundamental to understanding wind turbine design. Knowing this, you can quickly size up any wind machine by noting the dimensions of its rotor.

Relatively small increases in blade length or in rotor diameter produce a correspondingly large increase in swept area, and thus, in power. Compare the area swept by one wind turbine with a rotor diameter of 10 with that of another with a rotor diameter of 12. (Again, the units are not important here.)

$$A_2 = (R_2/R_1{}^2)\, A_1$$

$$A_2 = (^{12}\!/_{10})^2\, A_v = 1.44\, A_1$$

Increasing the rotor diameter by 20 percent (from 10 to 12) increases the capture area by 44 percent.

This exponential relationship between swept area and the power available also explains a crucial wind energy axiom: Nothing tells you more about a wind turbine's potential than rotor diameter. The wind turbine with the bigger rotor will almost invariably generate more electricity than a turbine with a smaller rotor, regardless of generator ratings.

In this book, listings of different wind turbines are ranked by swept area, not by

Rayleigh Wind Speed Frequency Distribution.

Frequency of Occurrence (%)

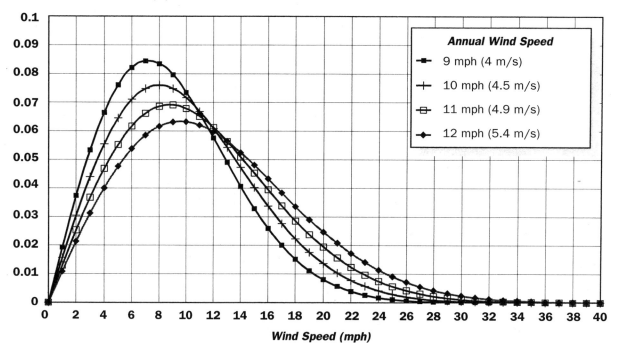

Chart plotting the frequency of different wind speeds at various annual average wind speeds for a Rayleigh distribution. The units are in percent of occurrence or the percent of time the wind will occur at that wind speed.

generator capacity. For example, Southwest Windpower's Air 303, which sweeps 1.02 square meters (m²), is listed as smaller than Aerocraft's AC 120 with a capture area of 1.13 m², even though the Air 303's generator is rated at 300 watts, or more than twice that of Aerocraft's 120 watts.

WIND SPEED DISTRIBUTION

Though we frequently speak in terms of power in referring to wind turbines, it's often just a substitute for energy. Energy is power over some unit of time. It's energy that we're after. It's energy in kilowatt-hours (kWh) that

we store in the batteries of an off-the-grid wind system, or energy in kWh that we sell back to the utility in an intertie system. Thus, we need to better define the wind speed used in the power equation, and we must account for how varying wind speeds affect the energy available.

We commonly designate the wind resource at a site by its average wind speed: typically, its average annual wind speed. (In off-the-grid battery charging systems, average monthly wind speeds are also commonly used.) However, because power is a cubic function of wind speed, periods of strong winds contribute far more to annual energy

Effect of Speed Distribution on Wind Power Density for Sites with Same Average Speed.

Site	Annual Average Wind Speed m/s	mph	Wind Power Density W/m²	Energy Pattern Factor or Cube Factor
Culebra, Puerto Rico	6.3	14	220	1.4
Tiana Beach, New York	6.3	14	285	1.9
San Gorgonio, California	6.3	14	365	2.4

Battelle PNL Wind Energy Resource Atlas 1986.

production than would be indicated by the average wind speed alone. Thus, the distribution of wind speeds over time is also important.

Jack Park, one of the pioneers in America's 1970s wind power revival, put it succinctly: "The average of the cubes is greater than the cube of the average." That is, the average of the cube of different wind speeds over time is greater than the cube of the average speed. To account for this, we need to know the actual distribution of wind speeds over time, assume a hypothetical distribution, or otherwise compensate with what Park has called the "cube factor" and others have labeled the "energy pattern factor."

The power in the wind at three different sites with exactly the same average wind speed illustrates the importance of Park's "cube factor." Though a New York site may experience the same average wind speed as one in Puerto Rico, the Caribbean island lies in the trade wind belt and has more constant winds. These steady winds produce less power over time than a temperate wind regime like that of New York. The blustery winds that rush through the San Gorgonio Pass contain 66 percent more power than the gentler winds bathing Puerto Rico.

Meteorologists have characterized the distribution of wind speeds for many of the world's wind regimes. For temperate climates such as that of the continental United States, the Rayleigh wind speed distribution offers a good approximation. Like New York's Tiana Beach, the cube factor for the Rayleigh distribution is 1.9.

WIND RESOURCES

Commercial wind power developers measure actual wind resources, in part, to determine the distribution of wind speeds because of the distribution's considerable influence on wind potential. Where this isn't practical, speed distributions at nearby meteorological stations may be used. If neither source is available, the Rayleigh distribution is often substituted.

Nor'Wester Energy Systems' Jason Edworthy says that small wind turbines seldom justify full-blown wind resource assessments, especially for off-the-grid use. In most stand-alone applications, the high value that a wind turbine adds to a hybrid wind and solar system warrants its use, often regardless of the wind resource. There are exceptions, of course. It makes no sense, for example, to install a wind turbine in a forest where trees will block the wind, just as it makes no sense to mount a solar panel under the shade of an awning.

Many purchasers of micro turbines have opted for the small machines over measuring the wind with a recording anemometer. Their reasoning is simple: Photovoltaic modules don't provide enough power in the winter and require a supplemental power source. A small wind turbine can be installed for about the same cost as a recording anemometer. While

the turbine is in use, its actual energy production can be tracked to quantify the wind potential.

This doesn't mean you shouldn't study the wind at your site. Monitoring the wind is instructive, says Edworthy. Hold a meter to the wind to develop a feel for its strength. Better yet, he says, fly a kite. Attach streamers to the line and watch how those streamers near the ground roil and flap, while those higher up smooth out. Those streamers tell you about something that's invisible—the wind. And where those streamers fly smoothly is where you want your wind turbine.

To locate sources of wind data search the Internet or, in the United States, contact the American Wind Energy Association or the National Renewable Energy Laboratory. In Germany, contact either the Deutches Windenergie Institut or the Institut für Solare Energieversorgungstechnik. In Great Britain, contact the Energy Technology Support Unit.

INCREASES IN WIND SPEED AND POWER WITH HEIGHT

Because obstructions near the ground disrupt the flow of the wind, wind speeds typically increase with height. Wind speeds can sometimes increase dramatically with height over rough terrain. This effect is so important that data on wind speeds will often include the height at which the wind was measured. If the height is not specifically mentioned, it is usually assumed to be about 10 meters (33 feet) above the ground. Most wind turbines will be installed on towers much taller than this to take advantage of the stronger, less turbulent winds aloft. To elevate a Bergey Excel above the trees in a shelter belt surrounding

Relative Size

In wind energy, size, especially rotor diameter, matters. Today, wind turbines range in size from the 20-watt Marlec 500, with a rotor only 0.5 meters (1.7 feet) in diameter, to Vestas's 1,650 kW giant, with a rotor spanning 63 meters (200 feet). There is no ironclad rule on what constitutes a small wind turbine. Size designations are somewhat arbitrary. Clearly the Marlec 500 is small, and the Vestas's 1.65 megawatt turbine is not.

Size classification depends upon both the diameter of the rotor and the capacity of the generator. Typically, small wind turbines encompass machines producing anywhere from a few watts to 10–20 kW. Wind turbines at the upper end of this range are driven by rotors 7–9 meters (23–29 feet) in diameter.

Small wind turbines can be subdivided further into micro wind turbines—the smallest of small turbines—mini wind turbines, and house-hold-size wind turbines. In this book we classify micro turbines as those from 0.5–1.25 meters (about 2–4 feet) in diameter. These machines include the 20-watt Marlec as well as the sleek 300-watt Air 303 from Southwest Windpower.

Mini wind turbines are slightly larger and span the range between the micro turbines and the bigger household-size machines. They vary in diameter from 1.25–2.75 meters (4–9 feet). Popular turbines in this category include World Power Technologies' H500 at 500 watts as well as Bergey Windpower's 850-watt model.

Household-size wind turbines (a translation of the Danish term *hustandmølle*) are the largest of the small wind turbine family. As you would expect wind turbines in this class span a wide spectrum. They include turbines as small as World Power's Whisper 1000 with a rotor 2.7 meters (9 feet) in diameter to the Bergey Excel that uses a rotor 7 meters (23 feet) in diameter and weighs in at more than 1,000 pounds (463 kilograms).

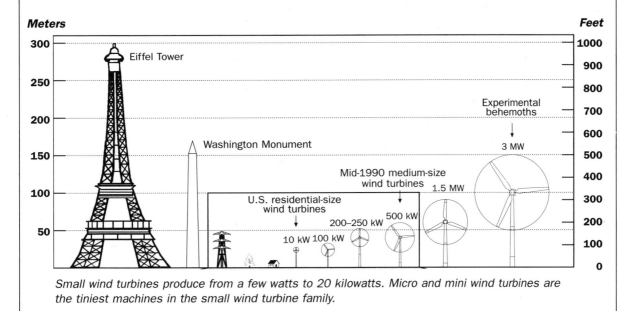

Small wind turbines produce from a few watts to 20 kilowatts. Micro and mini wind turbines are the tiniest machines in the small wind turbine family.

Increase in Wind Speed with Height.

$$V = V_o (H/H_o)^\alpha$$

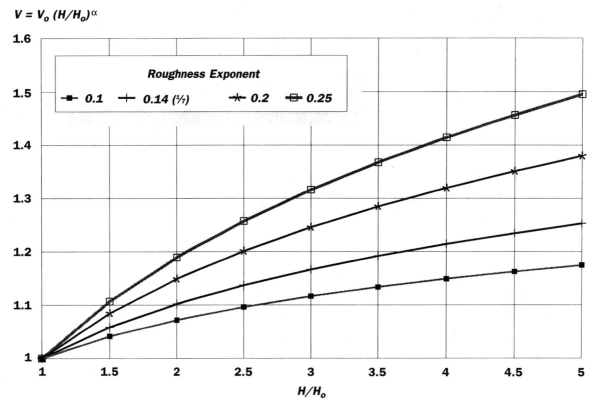

Simple chart for estimating the increase in wind speed as a function of relative tower height.

one Alberta home, Nor'Wester's Edworthy used a tower 100 feet (30 meters) tall.

The easiest way to calculate the increase in wind speed with height is to use the "power law" method. Another approach using logarithmic extrapolation is common in Europe. The power law equation may be less scientific, but it works well and is more conservative than the logarithmic method.

$$V/V_o = (H/H_o)^\alpha$$
$$V = (H/H_o)^\alpha V_o$$

Where V_o is the wind speed at the original height, V is the wind speed at the new height, H_o is the original height, H is the new height, and α is the surface roughness exponent.

The rate at which wind speeds increase with height varies with the vegetation, and the terrain. The increase is greatest over rough terrain or numerous obstacles, such as trees and shrubs, and smallest over smooth terrain, such as the surface of a lake. In the power law, this is reflected in the roughness exponent, α. Over smooth terrain it may be as low as 0.1, and over rough terrain as great as 0.25. For example, on farmsteads of the American Great Plains, the $1/7$ power law often applies. That is, the roughness exponent is $1/7$ or 0.14.

Consider the increase in wind speed when doubling tower height from 10 to 20 or from 30 to 60. (The units are unimportant; it's the ratio that counts.)

Increase in Wind Power with Height.

$$P = P_o (H/H_o)^{3\alpha}$$

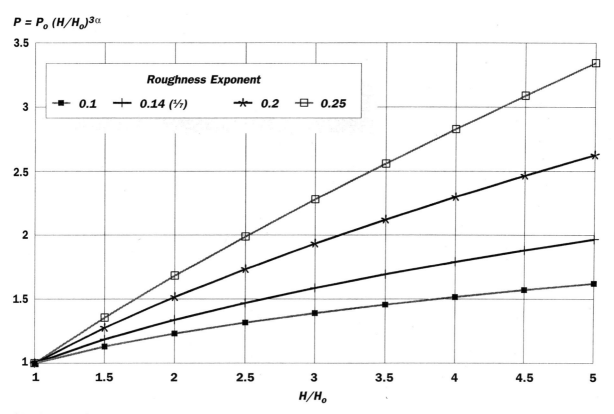

Simple chart for estimating the increase in wind power as a function of relative tower height.

$$V = \left(^{20}/_{10}\right)^{0.14} V_o = 2^{0.14} V_o = 1.1 \, V_o$$

On terrain where the $^1/_7$ power law applies, doubling tower height increases wind speed by 10 percent. Increasing tower height five times, say from 10 to 50 or from 30 to 150, may increase wind speed as much as 25 percent.

$$V = \left(^{50}/_{10}\right)^{0.14} V_o = 1.25 \, V_o$$

But power increases even more dramatically because of its cubic relationship with speed. Doubling tower height increases the power available by 34 percent.

$$P = (H/H_o)^{3\alpha} Po = (2)^{3(0.14)} P_o = 1.34 \, P_o$$

Increasing tower height five times nearly doubles the power available.

$$P = (5)^{3(0.14)} P_o = 1.97 \, P_o$$

This is why wind turbines are typically installed on tall towers.

Typical Roughness Exponents.

	Surface Roughness Exponent α
Water or Ice	0.1
Low Grass or Steppe	0.14
Rural with Obstacles	0.2
Suburb and Woodlands	0.25

Micro turbine on the beach. *A 50-watt Aerogen strapped to beach-front cabin on Denmark's Limfjord. The cabin is occupied only during the summer months. There is a 120-kilowatt wind farm turbine in the background.*

2

Estimating Performance

Be forewarned. While estimating how much energy a small wind turbine might produce at a given site involves some minor number crunching, the results are not as exact as in rocket science. Small wind turbines are notorious for defying expectations, especially in battery-charging systems. Even the experts have trouble. In fact, currently there is no international standard for how to measure small wind turbine performance. There are several reasons for this.

Unlike medium-size wind turbines that are linked to the utility grid, small wind turbines are used in a variety of applications. Wind turbines tied to the electric utility network feed an infinite sink. The utility system will consume all the energy the turbine produces. When the wind turbine captures the wind's energy and converts it to electricity, the turbine can deliver all of its generation to the grid. Thus, the electrical load on the wind turbine is predictable.

The situation with small battery-charging wind turbines is just the opposite: The load—the batteries—may not always be able to take the energy when it's available. When the batteries become fully charged, the turbine must spill, or dump, the excess energy that's available. Some manufacturers, such as World Power Technologies, offer dump loads, or circuits, for using this excess generation and keeping the wind generator fully loaded, but often small wind turbines simply spill the energy available in the wind when the batteries become fully charged. Thus, it's difficult to measure the performance of the small wind turbine because it may be spilling the energy in the wind instead of delivering it to a load.

Most small wind turbine manufacturers also lack the funds necessary to perform extended field tests on their products. These are small businesses with small staffs; they have little time or money for the painstaking tests needed to understand and evaluate their product's performance in the field. It was more than a decade after the introduction of the Bergey Excel before the company accumulated sufficient data to characterize the turbine accurately in battery-charging applications.

Relative Size of Small Wind Turbines.

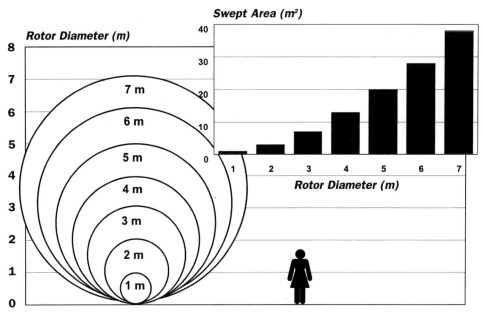

The swept area of some popular micro, mini, and household-size wind turbines.

With these caveats in mind, we can examine three methods for calculating the gross amount of energy small wind turbines may capture. The first uses the swept area of the rotor. The second uses the manufacturer's advertised power curve. The third approach simply uses the manufacturer's published estimates.

SWEPT AREA METHOD

Previously, we learned that rotor diameter—or more correctly, the rotor's swept area—is one of the critical factors in determining how much energy a wind turbine can capture. The other is wind speed. If we know how typical wind turbines work under ideal conditions, we can use rotor swept area and average annual wind speed to estimate annual energy output, or AEO.

Let's assume we want to use Bergey Windpower's 850 model. This mini wind turbine uses a rotor 2.4 meters (8 feet) in diameter and thus intercepts 4.5 m² (50 square feet) of the wind stream.

$$A = \pi R^2 = \pi(2.4/2)^2 = \pi(1.2)^2 = \pi(1.44) = 4.5$$

Assume also that we plan to install the turbine in the Texas Panhandle, where the average annual wind speed at hub height is 12 mph (5.4 m/s) and the distribution of wind speeds approximates a Rayleigh function. In the previous chapter we learned that a Raleigh distribution produces a cube factor of 1.9. With this information we can calculate how much wind energy the wind turbine will intercept. But we don't know yet how much of that wind energy it can capture.

The most aerodynamically sophisticated rotors can capture, at most, 40 percent of the energy in the wind. Generators, especially on small wind turbines, seldom convert more than 90 percent of the energy delivered to them. Throw additional losses into the equa-

Approximate Annual Energy Output of Small Wind Turbines.

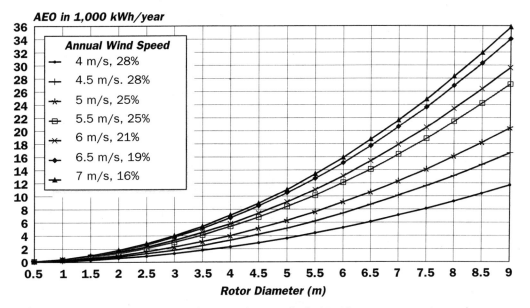

Chart for estimating the amount of energy that small wind turbines can generate yearly as a function of rotor diameter and wind resource. This chart summarizes estimates of AEO using rotor swept area and overall conversion efficiencies derived from performance estimates by dozens of different wind turbine manufacturers.

tion, to compensate for the turbine yawing in response to changes in wind direction, and you reach an overall conversion efficiency nearing 30 percent. You shouldn't expect more than this.

Despite all the hype in press reports about new airfoils, breakthroughs in generator technology, and so on, a small wind turbine can seldom deliver more than 30 percent of the energy in the wind, and then only under ideal conditions exactly matching the situation for which the turbine was designed.

The United States Department of Agriculture (USDA) tested a Bergey 1500 for five years in a wind-electric pumping application at its Bushland experiment station west of Amarillo, Texas. They found that the turbine was available for operation—the wind industry's measure of reliability—93 percent of the time. This is quite good for a small wind tur-

bine, though there are examples where small turbines have operated trouble-free for several years. The Bergey 1500 in this application converted 23 percent of the energy in the wind into electricity in winds from about 8 m/s (18 mph) to 11 m/s (25 mph). The annual efficiency over the turbine's entire operating range was much less.

Typical turbines will capture 20 percent or less of the energy available in the wind annually. Now, let's see what we can expect.

$$AEO = 0.00508 \; AV^3 \; (1.9 \; cube \; factor)$$
$$(8{,}760 \; h/yr) \; (20 \; percent) \; (1 \; kW/1{,}000 \; W)$$
$$= 0.00508 \; (50 \; ft^2) \; (12 \; mph)^3 \; (1.9)$$
$$(8{,}760 \; h/yr) \; (20 \; percent) \; (1 \; kW/1{,}000 \; W)$$
$$\approx 1{,}460 \; kWh/yr$$

Small wind turbines are typically designed to perform best in the low-wind regimes

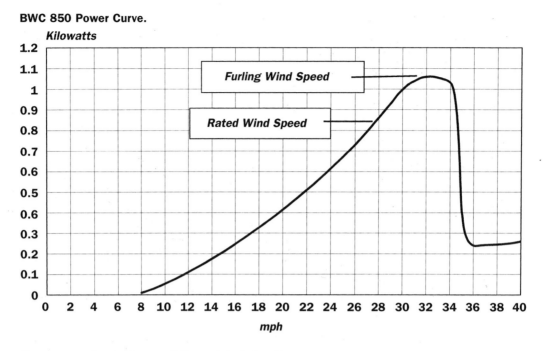

BWC 850 Power Curve.

Power curve for the Bergey 850, a mini wind turbine. Power is shown on the vertical axis, and wind speed is shown on the horizontal axis. Thus, if the wind is blowing at a certain speed at any one instant, you can find the corresponding power produced by the wind turbine at that speed. Note that the peak power exceeds the "rated" power and that power drops dramatically when the turbine is furled at wind speeds above 33 mph (14.7 m/s).

where most people live: sites with average wind speeds of 4–5 m/s (9–11 mph), for example. In locales with higher average annual wind speeds, their performance drops off dramatically. It's not uncommon at extremely windy sites for small wind turbines to convert only 10 percent of the energy in the wind. This is normal. Because of the cubic relationship with wind speed, there's so much energy available at windy sites that designers can afford to capture only a small part of it.

POWER CURVE METHOD

The second method for estimating annual performance is the same as that used by the manufacturers themselves when calculating what their wind turbine will do under various wind conditions. It uses the manufacturer's power curve and a distribution of wind speeds. A word of warning: The power curves proffered by some manufacturers of small turbines can best be characterized as informed guesswork. View power curves and the energy calculations that result with a good dose of skepticism. There are no government agencies ensuring the accuracy of published power curves, nor is there even agreement on how power curves for small turbines should be measured. It's still a black art.

All power curves include a start-up wind speed when the turbine begins spinning. Permanent-magnet alternators produce a voltage as soon as they begin turning. The voltage increases with increasing wind speed until the voltage needed for the application, such as battery charging, is reached. At this time the charge controller draws current from the alternator and power is produced. The cut-in wind speed occurs when the generator first begins producing power.

The power curve also includes the turbine's rated wind speed; that is, the wind speed at which the wind turbine produces the power indicated on its nameplate. Often, peak power is higher than rated power.

At a given speed, the wind turbine begins governing or limiting the power it produces. In the case of many small wind turbines, the rotor begins furling, or turning out of the wind. This reduces the power produced. The values shown in the power curve are typically for the turbine's output at hub height.

The Bergey 850 used in our previous example starts up at 8 mph (3.6 m/s) and reaches its rated power at 28 mph (12.5 m/s). The rotor furls, or folds, toward the tail vane at 33 mph (14.7 m/s).

If we know the distribution of wind speeds over time, we can match the speed distribution with the power curve and then sum the energy produced by the turbine across the range of wind speeds represented by the Rayleigh distribution.

At a wind speed of 18 mph (8 m/s), the Bergey 850 produces 327 watts. In a 12 mph wind regime with a Rayleigh distribution, an 18 mph (5.5 m/s) wind speed occurs 294 hours per year (see the graph in the previous chapter). In this speed bin, the turbine produces 0.33 kW x 294 h, or about 100 kWh per year. Across the entire speed range, we find that the Bergey 850 will generate about 1,400 kWh per year at this site.

MANUFACTURER'S ESTIMATES

If this has been more math than you want to deal with, most manufacturers provide

Calculating Energy Output for an Average Wind Speed of 12 mph.

To estimate annual energy output combine the power curve and the wind speed frequency distribution. In a 12 mph wind regime, a BWC 850 produces 0.3 kW in the 18 mph wind speed bin and will do so for about 3 percent of the time, or about 100 kWh per year.

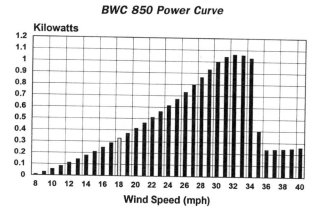

BWC 850 Power Curve

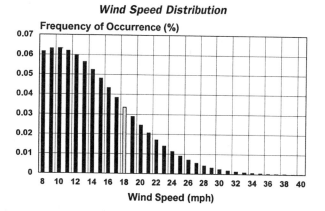

Wind Speed Distribution

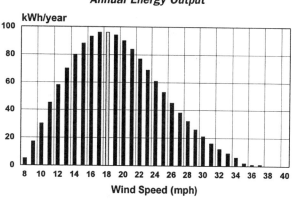

Annual Energy Output

Estimating Annual Energy Output for Bergey 850, Rayleigh Frequency Distribution.

Average Annual Wind Speed = 12 mph (5.5 m/s)

Wind Speed Bin (mph)	Instantaneous Power (kW)	Rayleigh Frequency Distribution*	Hours/Year	Energy (kWh/yr)
8	0.010	0.0616	539	5
9	0.030	0.0631	553	17
10	0.055	0.0632	554	30
11	0.082	0.0620	543	45
12	0.111	0.0597	523	58
13	0.142	0.0564	494	70
14	0.175	0.0524	459	80
15	0.210	0.0480	420	88
16	0.247	0.0432	378	93
17	0.286	0.0383	336	96
18	0.327	0.0335	294	96
19	0.370	0.0289	253	94
20	0.415	0.0246	216	90
21	0.462	0.0207	181	84
22	0.511	0.0171	150	77
23	0.562	0.0140	123	69
24	0.615	0.0113	99	61
25	0.671	0.0090	79	53
26	0.731	0.0071	62	45
27	0.795	0.0055	48	38
28	0.862	0.0042	37	32
29	0.930	0.0032	28	26
30	0.998	0.0024	21	21
31	1.040	0.0018	16	16
32	1.060	0.0013	11	12
33	1.055	0.0009	8	9
34	1.030	0.0007	6	6
35	0.400	0.0005	4	2
36	0.240	0.0003	3	1
37	0.250	0.0002	2	1
38	0.240	0.0002	1	0
39	0.250	0.0001	1	0
40	0.260	0.0001	1	0

Annual Energy Output 1,416

* If you have a computer and spreadsheet software, you can key in the formula for the Rayleigh distribution as ((@PI/2)*(cell for the wind speed bin/(cell for the average wind speed)^2)*@EXP(-(@PI/4)*(cell for the wind speed bin/cell for the average wind speed)^2)), where @PI and @EXP are software-dependent commands. If the 8 mph bin is A10 and the average annual wind speed is C9, then the cell C10 is (@PI/2)*($A10/(C\$9^2))*@EXP(-(@PI/4)*($A10/C\$9)^2) for Lotus 123 and its clones.

Manufacturer's AEO Estimate.

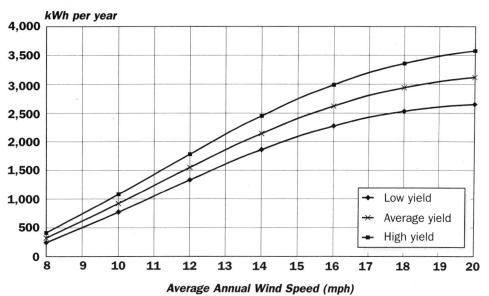

Bergey Windpower's estimate of annual energy production from its mini wind turbine, the BWC 850, in varying wind resources.

estimates of what to expect from their turbines under standard conditions, usually a Rayleigh distribution at sea level. The format varies. Some manufacturers provide a chart where AEO estimates are presented as a range of values; others display a single line. Presenting the data as a range of values emphasizes the uncertainty of the estimating process. For example, Bergey Windpower estimates their 850 model will produce between 1,400 and 1,600 kWh per year at an average wind speed of 12 mph (5.5 m/s).

Other manufacturers provide only one value, rather than a range of values. And suppliers of battery-charging wind turbines to the off-the-grid market often present this data as kWh per month rather than kWh per year. World Power Technologies, for example, estimates that its Whisper 500, which uses a 2.1 meter (7 foot) diameter rotor, will produce about 100 kWh per month at a site with a 12 mph average wind speed, or about 1,200 kWh per year.

No matter what technique is used, these projections represent the gross amount a turbine can be expected to produce. For a host of reasons, rarely is all this energy put to use. As mentioned previously, batteries that are fully charged may not be able to take more energy when, in a good stiff wind, the turbine is churning it out. Further, some of the energy that is eventually stored in the batteries is lost due to the inherent inefficiency of battery storage. Additional losses are incurred when you use an inverter to convert the DC stored in the batteries to run AC appliances. Maybe only 70 percent of the energy delivered to the batteries is eventually used productively.

Now that we have the tools to estimate the gross amount of energy that a wind turbine will produce, we can turn to the technology available today.

Old wind farm. *These three lonely 65-kilowatt WindMatics have watched over the Yellowstone River near Livingston, Montana, since the early 1980s. Note work platforms on the nacelles.*

3
Wind Turbine Technology

Wind turbines, especially small wind turbines, often confuse the uninitiated. Unlike photovoltaic panels, which typically all look alike, small wind turbines present a bewildering variety of shapes and sizes. At one time the situation was even worse, when practically every conceivable form was on the market. Fortunately, the technology has evolved toward a common configuration and though they may look different, most are actually quite similar. The differences today are more subtle, much like the differences between photovoltaic modules: differences in how they generate electricity and in how they are controlled. As Real Goods' Doug Pratt says, "Wind [energy] isn't something that's mysterious anymore."

CONFIGURATIONS

Wind turbines have been built in many different shapes; however, today nearly all small wind turbines are upwind, horizontal-axis turbines, where the rotor spins in front of the tower, about a line parallel with the horizon.

Two or Three Blades

There has been a long and bitter debate about the merits of using two or three rotor blades. (There was even a brief foray by one manufacturer of a turbine that used only one blade.) The only advantage of two blades over three is that two are cheaper. But it's a case of penny wise and pound foolish. Turbines with three blades run more smoothly than two, and that usually means they will last longer. After long experience with wind turbines, USDA's Nolan Clark opts for three-blade rotors over the two-blade machines still occasionally found.

Blade Materials

Most small wind turbines use composite materials, such as fiberglass (glass reinforced polyester), for their rotor blades. A few still use wood. Some have shifted to more exotic

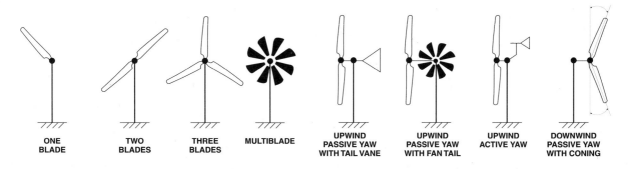

| ONE BLADE | TWO BLADES | THREE BLADES | MULTIBLADE | UPWIND PASSIVE YAW WITH TAIL VANE | UPWIND PASSIVE YAW WITH FAN TAIL | UPWIND ACTIVE YAW | DOWNWIND PASSIVE YAW WITH CONING |

Horizontal-axis wind turbines. *Though there are many different configurations of wind turbines, nearly all small wind turbines direct the rotor upwind of the tower with a tail vane. (Adapted from J.A. Twidell and A.D. Weir,* Renewable Energy Resources, *p. 211. Copyright 1986 by J.A. Twidell and A.D. Weir.)*

composites using carbon fiber instead of glass. None use aluminum because for its propensity for metal fatigue.

Orientation
Because of their size, small wind turbines can't accommodate the yaw motors and mechanical drives of the bigger upwind turbines. Nearly all small wind turbines use tail vanes to point the rotor into the wind. One of the few exceptions is Proven Engineering's downwind turbine, described later.

> Wisconsin's Mick Sagrillo is a proponent of what he calls the "heavy metal school" of small wind turbine design. Heavier, more massive turbines, he says, typically survive longer.

Robustness
Wind turbines work in a far more demanding environment than photovoltaic panels that sit quietly on your roof. You quickly appreciate this when you watch a small wind turbine struggling through a gale. There's no foolproof way to evaluate the robustness of small wind turbine designs. You certainly can't rely on the manufacturer's pronouncements. No manufacturer is going to tell you that their turbine is only suitable for light winds.

In general, heavier small wind turbines have proven more rugged and dependable than lightweight machines. Wisconsin's Mick Sagrillo is a proponent of what he calls the "heavy metal school" of small wind turbine design. Heavier, more massive turbines, he says, typically survive longer. "Heavier" in this sense is the weight or mass of the turbine relative to the area swept by the rotor. By this criteria, a turbine that has a relative mass of 10 kg/m^2 may be more robust than a turbine with a specific mass of 5 kg/m^2.

Overspeed Control
All wind turbines typically have a means for controlling the rotor in high winds. Overspeed control is one of the characteristics that sets different brands of wind turbines apart. Most micro and mini wind turbines furl, or

Wind Turbine Specific Mass.

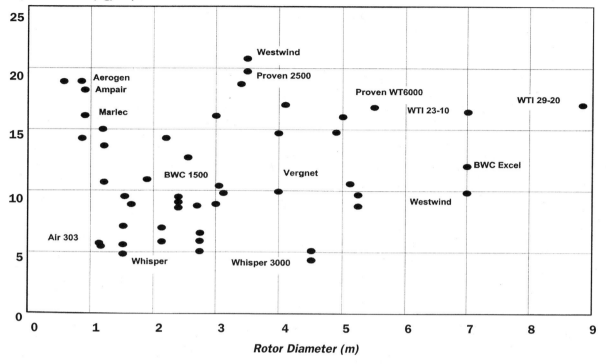

The weight of a wind turbine relative to the area swept by its rotor is a good indication of ruggedness. Wind turbines with a higher specific mass are generally more rugged than those with a lower specific mass. Typically turbines with a higher specific mass cost more than lightweight machines, but they last longer and operate more reliably.

fold about a hinge so that the rotor swings toward the tail vane. Some furl the rotor vertically (World Power's Whisper 500 and Southwest Windpower's Windseeker), others furl the rotor horizontally toward the tail (Bergey Windpower, Marlec, LVM, and World Power). Several designs of household-size turbines pitch the blades (Proven and Vergnet), and one pitches the blades and furls the rotor (Wind Turbine Industries).

To passively furl the rotor in high winds, the rotor axis must be offset from the furling axis. This is the tower or yaw axis for designs, like those of Bergey Windpower, that furl horizontally. In high winds, thrust on the rotor overcomes the restraining force keeping the rotor into the wind, and swings the rotor toward the tail vane. The wind speed at which furling occurs and the manner in which it operates is a function of the hinge between the tail vane and the nacelle, or body of the turbine, the design of which is far more complicated than most imagine.

The principle is the same for vertical furling in Southwest Windpower's Windseeker: In high winds the rotor tilts into the air, where it resembles a helicopter rotor. As in horizontal furling, the design of the hinge is critical to safe and smooth operation. Often there is a mechanism—a spring or shock

Horizontal furling. *Many small wind turbines protect themselves in high winds by furling or turning the rotor toward the tail vane: Note how the rotor axis of the Bergey Excel on the left is offset from the center of the tower (the yaw axis). This furling offset can also be seen on the Bergey 850 in the background.*

Horizontal furling. *Bergey 1500 furling toward its tail vane during high winds in the Texas Panhandle.*

absorber—to dampen the rate at which the rotor returns to its full running position.

Generators

Most small wind turbines use permanent-magnet alternators. This is the simplest and most robust generator configuration and is nearly ideal for micro and mini wind turbines. There is more diversity in household-size turbines. Bergey Windpower sticks with its permanent-magnet alternator, but Wind Turbine Industries uses a conventional wound-field alternator, while Vergnet uses an off-the-shelf induction generator.

One noticeable characteristic of some permanent-magnet alternators used by small wind turbine manufacturers, such as Bergey and World Power, is their inside-out design. The case to which the magnets are attached, sometimes called the magnet can, rotates outside the stator, or stationary part of the generator. In this configuration, the blades can be bolted directly to the case, and often are. There is also another benefit: Centrifugal force presses the magnets against the wall of the magnet can. In contrast, the magnets attached to the shaft of a more conventional shaft-driven alternator are thrown away from the spinning shaft. Because of the high rotational speeds found in small wind turbines, especially when unloaded, designers usually give special attention to retaining the magnets in shaft-driven alternators. For example, in the Air 303, Southwest Windpower straps the magnets to the rotor with a metal band.

Most small wind turbine alternators produce three-phase AC to make best use of the space inside the generator case. Some battery-charging models rectify the AC to DC at the generator; others rectify it at a controller that can be some distance from the generator.

Using permanent magnets poses a minor problem for small wind turbines in light winds when the rotor is at a standstill. Permanent magnet alternators suffer from cogging: The rotor shaft seems to stick when the magnets align with the stator coils, creating a resistance to motion between the two. Scoraig Wind Electric's Hugh Piggott notes that skewing the slots in the laminations of the armature reduces cogging and many small turbine manufacturers use this technique. The same effect can be created by skewing the magnets as seen in Southwest Windpower's Air 303.

MICRO TURBINES

With this background we can now look at some specific products, beginning with micro turbines. These are machines suitable for recreational vehicles, sail boats, fence charging, and other low-power uses. For example, micro turbines will generate about 300 kWh per year at sites with average wind speeds of 5.5 m/s (12 mph) like that found on the Great Plains.

This is an exciting new category of wind turbines for land-based applications. While not quite at the stage where you can simply put them up and walk away, micro turbines make wind energy easier than ever. "There's more maintenance in the batteries than in the turbine," says Real Goods' Doug Pratt. (See Selected Micro Turbines in Appendix 1.)

One of the challenges faced when discussing micro turbines—and small wind tur-

Vertical furling. *World Power's Whisper H1300 furling vertically in high winds at the Folkecenter for Renewable Energy's test field. Note that in this design when the rotor tilts up, the tail boom tilts down. This reduces the turbine's directional stability, causing the turbine to swing rapidly around the tower. World Power has discontinued this design.*

bines in general—is comparing apples to apples. There are no standards for small wind turbines, and this is painfully obvious in rating wind turbine output. Because wind turbines are designed for different markets and manufacturers have widely divergent views as how best to serve those markets, there is little consistency in power ratings.

Marlec Rutland. *Rutland 1800 uses dual disks for this version of their air-gap generator. Note furling tail vane.*

Marlec Rutland. *A redesigned 913 in operation (Marlec).*

When comparing wind turbines, always compare rotor diameter first. For example, Aerocraft's AC 120 intercepts 10 percent more area of the wind than Southwest Windpower's Air 303. If you must compare wind turbines on their power ratings, always consider the

---

When comparing wind turbines, always compare rotor diameter first.

---

wind speed at which the turbine is rated. The rated power for the Air 303 is much higher than that of Aerocraft's AC 120 because Southwest Windpower uses a higher rated wind speed. There is almost three times more power in the wind at 12.5 m/s (28 mph, the

rated wind speed of the Air 303) than at 9 m/s (20 mph, the rated speed of the AC 120). This difference in rated wind speeds easily accounts for the difference in the rated power of the two machines.

Only at the windiest locations do wind turbines operate for any length of time in wind speeds above 12 m/s (27 mph). Wind turbines installed in wind regimes where most people live spend most of their time operating in winds less than 12 m/s, generating less than rated power. Thus, a wind turbine with a power rating of only 100 watts could produce as much energy as a wind turbine with a rating of 300 watts when they both use the same size rotor even though it may appear that the 300 watt turbine will capture more energy.

LVM Aerogen

British manufacturer LVM builds a full line of micro and mini wind turbines. They have been building micro wind turbines under the Aerogen trade name with high power-density rare-earth magnets since 1985. Though most of LVM's multiblade turbines are destined for the marine market, some have also been used for other applications, such as electric fencing. LVM also builds low-voltage motors for pumping. There are a string of LVM's micro turbines marking the channel to Plymouth harbor in southwest England.

Marlec 913

Marlec Engineering's Rutland brand micro turbine can be found in yacht harbors around the world. The multiblade turbine is ubiquitous in small power applications in Great Britain, where frequent cloudy skies gives the feisty little machine an edge over solar panels. Like LVM, Marlec offers their turbines in land-based versions using a furling tail. Marine versions intended for yachting use a fixed tail, like the Ampair 100.

Marlec's Rutland 913 uses a novel pancake generator design that completely seals the generator in plastic. Hugh Piggott calls this an axial field generator, in contrast to the more common radial field designs used in automotive alternators or magnet-can alternators. Piggott offers a peek inside the Marlec in his book *Windpower Workshop* (Centre for Alternative Technology, 1997) and describes how the

Marlec Rutland. *Marlec's novel pancake or "air gap" generators being assembled at their plant in central England. This is all there is to Marlec's Rutland. Blades bolt to the plastic generator housing. The shaft is bolted to the nacelle frame.*

Ampair 100. *On rear deck of sailboat in Copenhagen's inner harbor.*

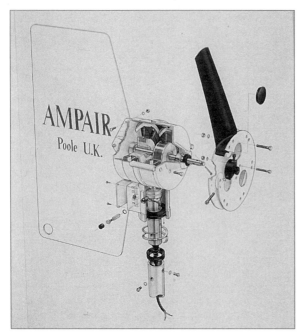

Ampair 100. *Exploded view (Ampair).*

unusual generator works. "Marlec windmills are cheap and efficient," says Piggott.

Ampair 100

If you see a small multiblade turbine in a yacht basin and it's not an LVM or a Marlec, it's most likely an Ampair 100. This rugged little machine uses a more conventional configuration than the Marlec. The shaft-driven alternator spins two, six-pole permanent-magnet rotors inside the cast aluminum body. The two six-pole stators are staggered 30 degrees to minimize cogging in low winds.

Southwest Windpower Air 303

Southwest Windpower put sex appeal into wind energy with the introduction of their sleek (some have said "swoopy") Air 303. Priced low to stimulate volume sales, the Air 303 is about half the price per rotor area of competitive products. The combination of price and visual appeal has worked, and Air 303s are finding wide use in low-power applications, from sail boats to telecommunications.

The lightweight Air 303 was intended as an off-the-shelf consumer commodity to complement photovoltaic panels in hybrid systems. Southwest Windpower hoped they would be used the same way as the solar panels, with off-the-grid users buying multiple units for low to moderate wind sites. Many were installed in windy regions, however, and this led to reliability problems.

The Air 303 has about half the specific mass of comparable turbines. This is partly due to the high performance airfoils that rely on blade flutter to limit power in high winds. Blades on competitive turbines are heavier. Consequently, the rotor and the entire turbine typically weigh less on the Air 303 than on competing products.

Another disadvantage of the Air 303's design was its built-in regulator. When it failed, and a number did, the turbines were repaired in the field or sent back to the shop. Similarly, one phase of the three-phase windings sometimes failed in the early units, says the Air 303's designer Dave Calley. This would leave the turbine running, but producing substantially less power than advertised. These problems have since been corrected, he says.

Southwest Windpower now offers a more rugged model for windy sites. Their "industrial" version sports cooling fins to keep the generator from overheating, greater clearance between the tips of the blades and the tower to eliminate blade strikes, and a choice between using a built-in regulator or not, to provide greater flexibility when charging batteries. The industrial version costs about twice as much as the standard model.

One clever trick to boost power, notes Marty Jones, an experimenter from Texas, is

Air 303. *Mounted on a gate post.*

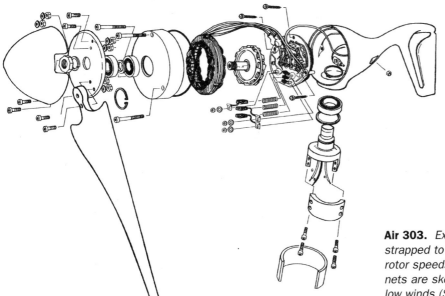

Air 303. *Exploded view. Note magnets strapped to rotor to retain them at high rotor speeds. Note also how the magnets are skewed to reduce cogging in low winds (Southwest Windpower).*

Whisper 600 (with "angle" governor). *World Power switched its Whisper 600 and H900 from vertical furling to horizontal furling (World Power Technologies).*

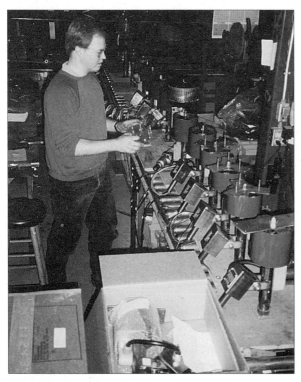

World Power. *Assembly of Whisper generators in Duluth, Minnesota (World Power Technologies).*

that the Air 303 uses three windings that are not identical. Each uses a different wire thickness and number of turns. In low winds where rotor speed is low, the turbine depends upon the winding using the thinner wire for production. At higher wind speeds, the windings with fewer turns of thicker wire can carry more current than the low-wind windings, delivering more power as it becomes available.

Despite anecdotal reports that the Air 303 can't live up to its power curve, tests at the National Renewable Energy Laboratory have shown that it does. Hugh Piggott found in his tests in Scotland that "it can thrash out 300 watts and more."

Aerocraft

This German manufacturer offers micro, mini, and household-size turbines from 120 watts to 5 kW. The two smallest units use vertical furling to control rotor speed, but the bigger models—from the AC 500 through the AC 5000—all use pitch weights to regulate the rotor. Like Bergey Windpower and World Power Technologies, Aerocraft's micro and mini wind turbines are designed for battery charging, and their household-size models for battery charging, heating, and interconnection with the utility network.

MINI TURBINES

Mini wind turbines are slightly larger than micro turbines, and are best suited for vacation cabins. They span the range from Southwest Windpower's Windseeker to Proven's WT600, turbines with rotors 1.5–2.6 m (5–8.4 ft), and include World Power Technologies' popular Whisper line. Wind turbines in this class can produce 1,000–2,000 kWh per year at 5.5 m/s (12 mph) sites. See Selected Mini Turbines in the *Resources* section.

World Power Technologies

World Power builds a line of turbines from 500 W–4.5 kW at its shop on the shores of Lake Superior. They offer each product in two versions: a standard two-blade model, and a three-blade high wind version. The high wind version, or H series, is essentially the same turbine as the standard series, says World Power founder Elliott Bayly, except that the turbines are rated for higher wind speeds and power output. On the H series, the addition of one blade (from two to three blades) increases the weight of the rotor slightly. The heavier rotor stays pointed into the wind longer before it begins to furl, enabling it to capture more energy in high winds than the two-bladed version. They rewind the generator to handle the greater power that results. World Power's three-blade rotors operate more smoothly than their two-blade cousins.

In 1997 World Power introduced new airfoils, blade construction, and a new furling mechanism. The smaller turbines (the H500, 600, and H900 models) use injection-molded, polycarbonate blades with fiberglass reinforcement. The blades are torsionally stiffer, more moisture resistant, and less prone to wind erosion on their leading edges than the wooden blades used previously. On the bigger turbines (the 1000, H1500, 3000, and H4500), World Power now uses carbon fiber reinforcement in its composite blades.

World Power has also switched from the vertical furling, once characteristic of its turbines, to a new "angle" governor. The conversion of Whispers from the old vertical furling

Bergey 850. *Karl Bergey holding magnet can to BWC 850 with magnets glued inside.*

Bergey 850. *Close up view of magnet can (rotor), armature, tail vane hinge, and yaw assembly; pen for scale.*

to the new horizontal furling enables better control of the turbine in high winds than does the bobbing action of the previous design.

Bergey Windpower

In 1994, Bergey Windpower launched the smallest turbine in its line: the BWC 850. Like World Power, Bergey Windpower bolts the blades to the generator case and spins the magnet can around the stator. Bergey Windpower uses an unusual rotor design in which the blades are set at a high angle of attack for starting the turbine in low winds. As rotor speed increases, a weight twists the torsionally flexible, pulltruded fiberglass blades. This passively and progressively changes blade pitch toward the optimum running position. As rotor speed increases, centrifugal force stiffens the slender blades.

Bergey's 850 is a scaled-down version of their successful household-size turbines. It even uses the same blades as the Bergey 1500, though shortened for the smaller rotor. Like those of its bigger brothers, the Bergey 850's rotor furls horizontally toward the tail vane.

HOUSEHOLD-SIZE TURBINES

As their name implies, wind turbines in this class are suitable for homes, farms, ranches, small businesses, and telecommunications. They span an even broader range than other size classes, from as low as 1 kW, to more than 20 kW. Household-size wind turbines can generate from 2,000 kWh to 20,000 kWh per year at 5.5 m/s (12 mph) sites, like those on the Great Plains. (See Selected Household-size Turbines in Appendix 1.)

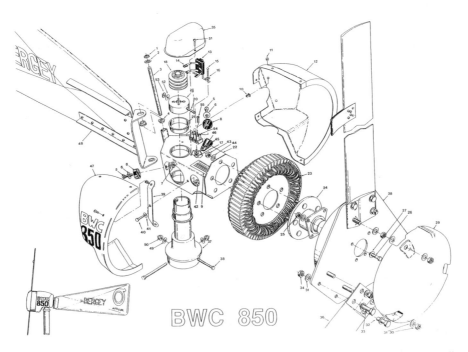

Bergey 850. *Exploded view. Note wheel bearing, stator coil, blade pitch weight, and furling tail vane. Permanent magnets (not visible) are mounted inside the rotor housing (Bergey Windpower).*

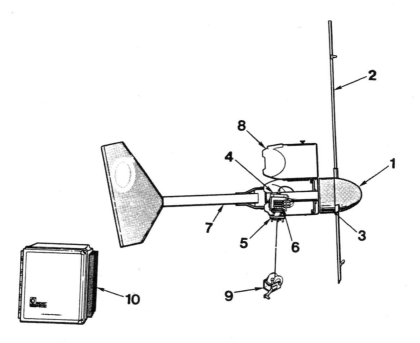

Bergey 1500. *1. Spinner or nose cone, 2. pulltruded fiberglass blades, 3. permanent magnet alternator, 4. mainframe, 5. yaw bearing, 6. slip rings and brushes, 7. tail vane, 8. nacelle cover, 9. winch for furling rotor out of wind, 10. control panel (Bergey Windpower).*

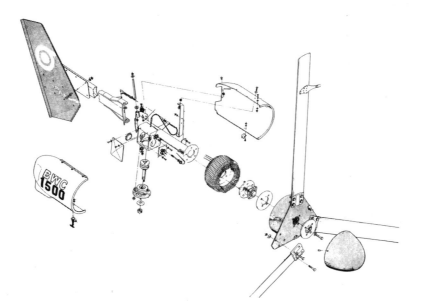

Bergey 1500. *Exploded view. Note wheel bearing that supports rotor, stator or armature coil, and hinged tail boom (Bergey Windpower).*

World Power's Whisper 1000 and H1500, Bergey Windpower's 1500 and Excel, and Wind Turbine Industries' 9-meter (29-foot) turbine are probably the most wel-known products of this class in North America. But in the South Pacific, Westwind is equally well known; in Great Britain, Proven Engineering has been building a reputation; and in the French-speaking world, Vergnet dominates the market.

Proven Engineering

Wind turbines have to be tough to withstand the fierce gales on Scotland's northwest coast where Hugh Piggott and his family live. Piggott, who began working with wind energy by building small turbines for his neighbors on Scotland's Scoraig peninsula, designs, tests, and services small wind turbines. His neighbors "still ring me up when they need their windmill fixed," says Piggott. With thirty small wind turbines of various sorts within a short walk of his home, Piggott is the authority on small turbines in Great Britain.

At Scoraig Wind Electric, Piggott has installed several of Proven Engineering's turbines in remote corners of Scotland. One had to be carried over a mountain to a youth hostel. Another was installed at a railway station accessible only by rail or footpath. "It was a breath of fresh air to work with the Proven machines," says Piggott. "They are built like steam trains. These are windmills which can just stand there on a windy headland and whack out kilowatts nonstop for years."

The Proven machines—like those of World Power and Bergey—also use a direct-drive, permanent-magnet alternator. There the similarity ends. The most obvious difference is Proven's downwind rotor. Without a tail vane, Proven can't rely on furling to limit rotor speed in high winds. They use an intriguing flexible hinge between the blades and the hub that allows the blades to change pitch with rotor speed. Like all downwind turbines, the Proven machines will sometimes precess, or walk, around the tower, occasionally running upwind.

Wind Turbine Industries

Wind Turbine Industries' turbines are the largest machines listed in this book. These machines can be used in utility intertie applications. Wind Turbine Industries uses designs, patterns, and parts from the defunct Minneapolis-based manufacturer Jacobs Wind Energy Systems. About half of the parts for each turbine come from inventory remaining after Jacobs closed shop, says WTI's Steve Turek.

These turbines should not be confused with 1930s-era "Jakes." Though the design looks like it's from the 1930s, it's not. The two designs are radically different. The 23-foot (7-meter) model drives a 17.5 kW alternator mounted vertically in the tower via a hypoid gearbox. The 1100 rpm, four-pole alternator produces 40 hertz, three-phase AC, which is then rectified and inverted before delivery to the utility. Though the turbine once used wooden blades, WTI has replaced these with fiberglass blades made from a resin transfer process.

NOVEL TURBINES

Occasionally a company announces a breakthrough in wind turbine development that will make all competing wind turbines obsolete. Inventors, and many engineers as well, often can't see the forest for the trees and get swept along by their enthusiasm. They issue

1980s Jacobs. *Household-size wind turbine that uses blade-actuated governor and right-angle drive for tower-mounted alternator.*

press releases, garner a few articles in local newspapers, raise money from hapless investors, build a prototype or two, find that their dream machine doesn't work quite the way they expected, and then sink back into oblivion.

This scenario happens frequently with concentrator designs. These are wind turbines with funnels or shrouds that either concentrate or augment the flow through the rotor. These turbines have never operated as touted, and are quickly forgotten except by those who lost money in the ill-advised ventures. The reason these designs don't work is simple: Building ducts or shrouds around

wind turbines is rarely justified because it's almost always easier to extend the blades to compensate for any speed-up effect that a concentrator may produce.

Even promising designs from legitimate manufacturers encounter enough technical and financial problems that only a few survive. Darrieus, or "eggbeater" turbines, are one example. Sleek and relatively simple, they promised lower costs and greater reliability than conventional wind turbines. But they didn't deliver. Darrieus designs failed first in the demanding small turbine market, then in the commercial or wind farm market. Today only a few Darrieus turbines remain

Darrieus wind turbine. *Practically no one is working with vertical axis wind turbines today. Here the Alternative Energy Institute's Vaughn Nelson demonstrates how a small Darrieus turbine can be used to pump water near Amarillo, Texas, in the late 1970s.*

Experimental 1.5-kilowatt battery charger. *The 2.7-meter Windflower uses a multiblade rotor to drive a standard induction generator via a readily available gearbox. The induction generator enables Windmission to regulate the generator's field. In high winds, the rotor furls toward the tail (Windmission).*

standing, and soon these will be removed for scrap.

But there is still a need for better small wind turbines—designs that are both less costly and yet more reliable than those presently on the market. And despite the risks, designers continue to experiment with new products.

TOWERS

Small wind turbines are installed on free-standing lattice towers similar to those of tra-ditional farm windmills, free-standing tubular towers, or guyed masts. Tubular towers may be either straight-walled or tapered. Guyed masts use either lattice tower sections, pipe, or tubing depending upon the design.

Tapered tubular towers are the most visually pleasing but also the most costly, and unless hinged for lowering to the ground are also the most difficult to climb for servicing the turbine. Free-standing lattice towers are nearly as expensive.

Guyed towers are the least costly. If they are hinged, guyed towers can be raised and lowered for servicing the turbine. Bergey Windpower recommends a guyed lattice mast for its household-size turbines. These machines are heavy and awkward, so that raising and lowering them is both risky and time consuming. For minor repairs, it's simpler to climb the latticework to the top of the tower than to lower the entire mast and turbine to the ground.

The recent boom in micro turbines is partly due to the design of inexpensive, lightweight mast systems. These tower kits use metal tubing engineered specifically for popular micro and mini wind turbines such as those of Bergey Windpower, Southwest Windpower, and World Power Technologies.

Rooftop Mounting

Rooftop mounting of wind turbines, no matter how small, remains controversial. Southwest Windpower first suggested installing micro turbines on rooftops when they launched their Air 303. Their intent was to compete with the simplicity of mounting photovoltaic panels. They recommend installing the Air 303 on "as tall a tower as possible," and offer both 25-foot (7.5 meter) and 47-foot (14 meter) towers. Southwest continues to suggest rooftop mounting as an inexpensive alternative, though they now stress that the building should be unoccupied.

All wind turbines vibrate and transmit the vibration to the structures on which they're mounted. All rooftops create power-robbing turbulence that shortens a wind turbine's life. Even if you were able to design a sophisticated dampening system that isolated the wind turbine from the structure, you couldn't avoid the turbulence. In the long run, it's not worth the risk or the trouble.

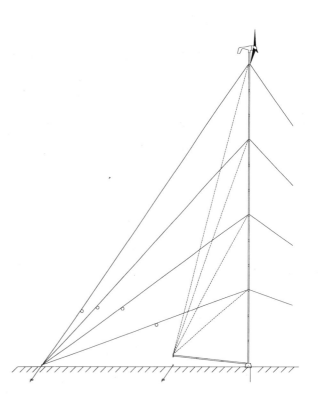

Tilt-up tower. *Lightweight tubular masts of guyed metal tubing packaged with gin pole and screw anchors offer an ideal choice for micro turbines. Note gin pole at right angle to vertical mast (NRG Systems).*

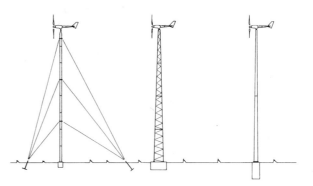

Tower types. *Guyed mast (left), free-standing truss or lattice tower (center), free-standing or cantilevered tubular tower (right) (Bergey Windpower).*

High-performance prototype. *University of Newcastle's experimental 5-kilowatt turbine at Fort Scratchley, New South Wales, Australia. Note unusual forked tail vane and, in the background, the fleet of ships waiting to load coal for export.*

4

Off-the-Grid Applications

wenty years ago, we envisioned the development of wind energy in the United States as an extended network of small wind turbines interconnected with the local utility. Every farm, ranch, and rural home would have its own. Each year, manufacturers would produce thousands of wind machines destined for utility intertie. But it hasn't worked out that way.

In parts of Denmark and northern Germany, nearly every farm has its own turbine. But these are large, commercial-size wind turbines, many producing more than one million kilowatt-hours each year. Excess generation is sold to the utility, just as milk is sold to the local cooperative.

In North America, the boom in small wind turbines has been in battery-charging systems. The reasons are part political, part technological. In the United States, utility intertie simply doesn't pay. But equally important has been the development of inexpensive micro and mini turbines that make wind energy affordable and easy to use, and

the development of new inverters, compact fluorescent lamps, low-power electronic devices, and photovoltaic modules that allow many to live off the grid.

The photovoltaic industry has aggressively sought out and serves the needs of a growing number of off-the-grid applications. The spin-off has been increasing interest in small wind machines, notably micro and mini turbines, for supplementing existing solar systems. This improves off-the-grid performance in winter, when solar's contribution is at a minimum. Elliott Bayly says 80 percent of World Power's turbines are used in hybrid systems alongside photovoltaic modules. Bill Dorsett, of Kansas, adds that "throughout the mountainous West there are isolated valleys where the cost advantages of powering remote cabins with solar cells are so compelling that no one thinks twice. Today there are livestock pumps, fence chargers, and certainly whole farmsteads that are remote enough to be cost-effective for stand-alone solar electric or wind power."

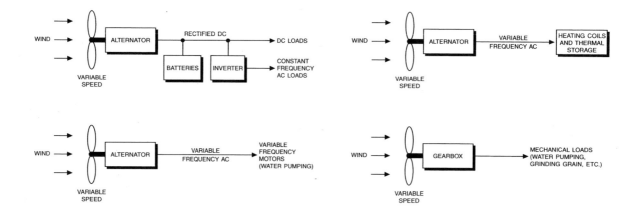

Stand-alone wind machines. *Small wind turbines can be used off-the-grid in a number of applications. Commonly they are used to charge batteries for providing both AC and DC loads (upper left). With modern electronic controllers, they can also drive electric well pumps and other motors directly without batteries (lower left). Small wind turbines can also be used to heat homes (upper right). In much of the world, remote homesteads and ranches still use farm windmills to mechanically pump water (lower right).*

HYBRID WIND AND SOLAR SYSTEMS

You might say that joining wind and solar power together is a marriage made in heaven. The two resources and technologies are complementary. Together, they not only improve the reliability of a stand-alone power system, but also are more cost-effective than either one alone.

Many off-the-grid homes start with a few photovoltaic panels, because they are simple to install and their unit costs are within reach of most homeowners. A small system may use two or more panels, either in series or parallel, a few batteries, and some 12- or 24-volt DC appliances.

Until recently, adding a wind turbine to the mix was problematic because even the smallest turbines cost more than most people were able or willing to spend. However, the advent of inexpensive micro turbines has lowered the cost of building a hybrid wind and solar system. The wind turbine is usually

not more than 50 percent of a hybrid system's total cost, says World Power's Eliott Bayly.

Because wind has higher power density than solar, even at low-wind sites, the addition of a small amount of wind capacity, such as that provided by a micro turbine, can significantly boost the total energy available.

Components such as batteries and inverters are critical to the success of hybrid systems. As hybrid systems have become more sophisticated, so too have the various components used in their design.

Batteries have always been an expensive and troublesome part of off-the-grid systems. Consider that a typical 6-volt storage battery has a gross capacity of 200 amp-hours, equivalent to about 1 kWh of chemical energy, and costs nearly $100. Thus, batteries cost about $100 per kWh of gross capacity, not counting freight. And shipping heavy batteries is costly. Moreover, not much more than 50 percent of the energy stored in a battery can be withdrawn without sulfating the plates and

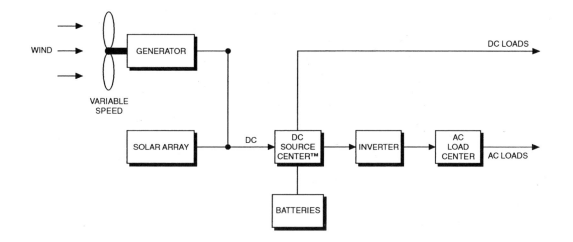

Small vacation home or cabin. *Cottage-size hybrid systems often use 12 or 24 volts with a micro turbine, one or two solar panels, two or more batteries, and sometimes an inverter for small AC loads.*

reducing its effectiveness. Batteries also have a limited lifetime. The Folkecenter for Renewable Energy estimates that batteries are good for about 2,000 cycles. (Batteries are still usable after 2,000 cycles, but they have reduced capacity.) If a battery discharges 50 percent of its gross capacity through 2,000 cycles, it will deliver about 1,000 kWh of net electrical energy over its operating lifetime. Thus, battery storage alone costs more than $0.10 per net kilowatt-hour of usable energy in an off-the-grid system ($100/1,000 kWh). Because of such high costs, designers make every effort to limit the amount of battery storage needed, by using the sun and wind together.

For recreational vehicles, cabins, or Third World systems, don't overlook the 220 amp-hour golf cart battery, says World Power's Bayly. These batteries are so widely available "they've become a world standard," he says. The 350 amp-hour L16 batteries from Trojan are also popular, but half again more expensive than the golf cart batteries.

For larger systems with AC loads, invert-ers are a must. Today's electronic inverters have made living off the grid with AC appliances easier than ever before. Real Goods' Doug Pratt says that new inverters, such as Trace's SW (Sine Wave) series, can be programmed to start certain heavy loads when excess power is available, cut the load when battery voltage falls, or start and stop a backup generator as needed.

Integrating the various components used in hybrid systems can be made easier by using a DC Source Center™. This is a convenient enclosure where the battery, wind turbine, and inverter connections can be made to the DC bus bar and safely fused.

Folkecenter Comparison of Generation.

	m^2	kWh/yr	Max kWh/mo	Min kWh/mo
Solar	1	100	12	2.2
Wind	1	200	22	12

Wind: 4 m/s (9 mph) average annual wind speed.

Source: Folkecenter for Renewable Energy.

Wind and solar hybrid. *An off-the-grid hybrid wind and solar system at Real Goods' Solar Living Center in Hopland, California. World Power Whisper 3000 on a do-it-yourself, tilt-up tower.*

Tracking photovoltaic modules. *Solar cells are an essential part of any hybrid power system.*

Mike LeBeau

Michael LeBeau's home, north of Duluth, Minnesota, is a good example of how various components are integrated into a wind-solar hybrid. LeBeau installed a Whisper 1000 with carbon fiber blades on a 65-foot (20-meter) tilt-up tower. He built the guyed tower himself from steel pipe.

Like many others, LeBeau began living off the grid using only solar energy. For satisfactory performance in the depth of winter in the North Woods, 2 kW of solar panels were necessary.

LeBeau's system also includes twenty Trojan L16 heavy-duty batteries, a Trace SW 4024, and a 2.2 kW Honda gasoline generator. LeBeau figures he's invested $15,000 in his home-built, 24-volt system. That's substantially less than the $25–$30,000 that it might have cost, says LeBeau, if he'd had a complete system designed and installed by a contractor. The local utility would have charged $10,000 to extend the line to his remote home overlooking Lake Superior. But the work and the investment were worth it to him, partly because the utility would have had to clear a swath of trees to his place to install an ugly line of poles.

Prior to adding wind power to their system, LeBeau's family adjusted their lifestyle to the availability of solar energy, using few appliances during the winter. Now, with their Whisper 1000 on line, LeBeau doesn't have to run his backup

Mike LeBeau's Whisper 1000. *LeBeau uses a Whisper 1000 and a 2-kilowatt fixed-array of photovoltaics to power his off-the-grid home outside Duluth, Minnesota.*

generator as much in winter as before. Should he generate excess power, he's programmed his Trace inverter to dump the surplus into an electric water heater.

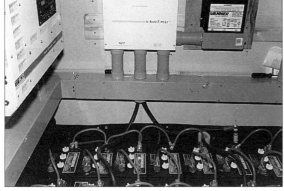

Mike LeBeau's battery room. *Trace 4024 inverter (left), Ananda Power Technologies power center with disconnect switch (center); Vanner charge equalizer (right); Interstate batteries (bottom). Note that all cables run in conduit or raceways. Ideally, the batteries should be isolated from the electronics in a well-ventilated room, but this installation is clean, tidy, and uncluttered.*

WIND PUMPING

For centuries, wind energy has been used to pump water. There are more than one million water-pumping windmills still in use worldwide. But today, wind technology offers more options for pumping water than it did just a few years ago: traditional mechanical wind pumps (farm windmills), air pumps, and wind-electric pumping. There are advantages to each.

The classic farm windmill "is hard to beat in light wind regimes," says Eric Eggleston, a researcher at the USDA's experiment station in Bushland, Texas. The farm windmill's high-torque rotor was designed for pumping in the light winds of summer on America's Great Plains, and does the job well. But, whether a farm windmill is an optimum choice or not depends upon the wind available, the depth to water, and the amount of water needed.

Though an American farm windmill costs about 10 percent less than an equivalent size wind-electric system, explains Eggleston, the farm windmill will only pump half the volume. This is due to the better aerodynamic performance of the rotor on the modern wind-electric turbine, and to a better match between the rotor's performance and the power available in varying winds.

The siting of mechanical wind pumps is also limited. Farm windmills must be placed directly over the well, whereas wind-electric pumping systems permit the wind turbine to be placed to best advantage. An electrical cable is then used to connect the wind-electric turbine with a pump motor at the well. For low-volume applications, air-lift pumps also offer similar flexibility, since they use pliable plastic tubing.

Unlike earlier designs with batteries and inverters, contemporary wind-electric pumping systems drive well motors directly. The key has been development of electronic controls that match the pump motor load to the power available at different wind speeds. The control system is the limiting factor, says Nolan Clark, director of the USDA's Bushland station. Newer technology should eventually cut the cost of controllers in half, making wind-electric pumping even more attractive.

As an example of what wind-electric pumping can do, Clark notes that one West Texas farmer uses two Bergey Excels for irrigating cotton. Each turbine pumps 20 gallons per minute (0.076 m³/min) from 300-foot (100-meter) wells, when the wind is available.

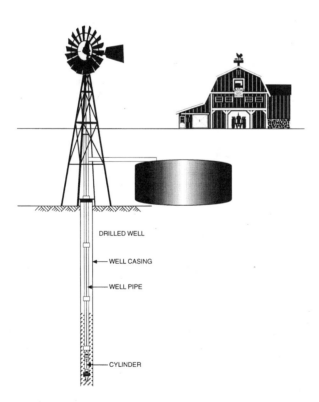

DRILLED WELL

← WELL CASING

← WELL PIPE

← CYLINDER

Wind pumping. *Historically, wind energy has been most often used to pump water.*

Large diameter mechanical wind pump. *25-foot (7.6-meter) diameter farm windmill for livestock watering at remote cattle stations in the Australian outback (Southern Cross).*

Air-lift or bubble pumps. *Bubbles of air created by these mechanical wind turbines can be used to lift water in wells or aerate farm ponds during the winter: (left) Bowjon air-lift pump in California; (right) Koenders wind pump in Alberta, Canada.*

WIND HEATING

Like wind and solar hybrids, using winter winds to heat your home has always seemed an ideal way to marry a technology with a natural cycle. Since heating loads are a function of heat-robbing winds, why not use those very winds to heat your house? This can be done quite simply by using a wind generator to power electric heating elements. The University of Massachusetts proposed such a wind furnace in the mid-1970s, and several companies have tried to market the idea to homeowners already on the grid. The concept never caught on in North America, where the economics never made sense, but it did in Denmark, where heating prices are considerably higher.

Denmark's Folkecenter for Renewable Energy has found that a wind turbine that covers winter heating demands can easily cover domestic hot water loads in summer.

Ed Wulf

When Ed Wulf was building his home in Southern California's Tehachapi Mountains, the local utility offered to bring him power for a mere $50,000. He told them "I can do better than that." He only had to look out his picture window for the solution: His window opens on to one of the world's largest wind power plants. The five thousand wind turbines across from Wulf's homestead churn out enough electricity to serve the needs of five hundred thousand Californians. For regulatory and logistical reasons, the area's wind power plants were unable to help Wulf. He had to go it alone. But the very existence of the nearby turbines proved that wind energy would work for him.

Wulf set out to install his own stand-alone power system, a hybrid that would use the area's wind and solar energy. Now Wulf's wind turbine attracts nearly as much attention as the big machines on the nearby hillsides.

Wulf installed a Bergey Windpower 1500, a large photovoltaics array, batteries, a diesel generator, and an inverter to power his conventional home. His was not a minimalist approach. Wulf's system could probably power an entire Third World village. But it works: It pumps water from a deep well, drives an evaporative cooler during California's hot summer months, and powers all the appliances common in an American household.

Household-size hybrid wind systems. *Ed Wulf's Bergey 1500 on 80-foot (24-meter) guyed tower near Oak Creek Pass in the Tehachapi Mountains. Though his front window opens on to one of the world's biggest wind power plants, Ed Wulf's homestead is off the grid and powered by his own wind and solar system.*

Home heating, old. *SJ Windpower's 6-meter (20-foot) diameter windrose provides electrical heat for a home (not visible on the left) in windy northwestern Denmark. Though this 10-kilowatt turbine is no longer manufactured, there are 30 still operating in Denmark. Medium-size, grid-connected wind turbine in the background.*

The Folkecenter has also found that it is economically more advantageous to use wind-generated electricity as electricity, rather than converting it to heat. Typically, electricity has double the value per kWh than does heat.

Proponents have argued that storing excess wind energy as heat is much cheaper than storing it in batteries. Thus heating with wind can play an important role in remote stand-alone systems that don't have ready access to fossil fuel for heating. Scoraig Wind Electric's Hugh Piggott dumps his excess wind power into "storage heaters." These are common in Great Britain for storing cheap, off-peak electricity at night for those on the grid, says Piggott, and are readily available for off-the-grid systems.

To guarantee a reliable supply of wind-generated electricity for a remote power system, you need a much bigger turbine than would be necessary to produce the appropriate number of kWh for a grid-connected home. The bigger turbine eases the load on the batteries by generating current even in light winds. In moderate or strong winds, the turbine produces more power than needed. The surplus can then be used for heating, thus saving fuel.

Dumping or diverting excess wind power to heating has become a common practice in stand-alone wind systems. For those off the grid, the situation in Alberta, Canada, is the same as in Scotland. In Alberta, Jason Edworthy used a Bergey Excel in a battery-charging system where excess energy was diverted to a conventional hydronic system for in-floor heating.

The turbine's electrical control system must be adapted to direct excess power to the dump load to ensure that the dump or diversion load is compatible with the wind turbine. If not, the wind turbine's rotor may stall or stop when the dump load switches on. Modern solid-state electronics are well suited for this function, and can be used to adjust the heating loads to the power available from the wind turbine.

Home heating, new. *In the belief that there's a demand for wind-powered home heating, Calorius has introduced this 5-meter (16-foot) diameter turbine in Denmark. The Calorius uses a water churn and delivers hot water to home hydronic (radiator) heating systems. Bugatti 57S for scale. Note guyed tilt-up tower with gin pole (Calorius).*

Micro wind turbines for recreational vehicle. *Two Air 303s atop a bus converted to a motor home. Solar panels in this hybrid system are mounted flush with the bus roof. Micro wind turbines are found throughout the marine and RV industry (Nancy Nies).*

Vacation homes. *LVM's Aerogen micro turbine powers small loads at this vacation cottage in a remote corner of northwestern Denmark. The junction box strapped to the pipe tower below the turbine both supports the power cable leading down the tower and protects the connections between leads from the turbine and the power cable.*

RECREATIONAL VEHICLES

In addition to traditional uses, such as pumping water and powering remote homesteads, today's small wind turbines are finding an increasing number of new applications. Thousands of micro turbines are used on sailboats around the world, and now they're appearing strapped to the sides of the lumbering land yachts that cruise America's byways. Since recreational vehicles (RVs) already have batteries and a charging circuit in place, micro turbines are an ideal addition to the 12-volt RV system. Many RVs sport solar panels to avoid running their noisy generators; micro turbines are a natural adjunct.

CABINS AND COTTAGES

It's a simple step from using a wind turbine on an RV to installing a micro turbine and batteries in a cabin or small vacation cottage. These 12-volt systems often use the same hardware and appliances found in RVs.

Electric fence charging. *This 12-volt Marlec 910F on a guyed pipe tower is charging an electric fence on the banks of Denmark's Skibsted Fjord. Solar-powered electric fences are common in Europe, but micro wind turbines are gaining ground.*

Telecommunications. *Household-size wind turbine (Bergey Excel) powering a QuebecTel mobile telephone station at Point-au-Père on the St. Lawrence River, Canada.*

Electric vehicles. *Household-size wind turbines can be used to charge electric vehicle batteries. Nancy Nies driving the Folkecenter for Renewable Energy's EV in northwest Jutland. The Danish slogan says, "Drive like a dream on windmill current."*

Cottages with heavier loads than those on RVs may opt for a more powerful 24-volt system.

ELECTRIC FENCE CHARGING

Once the sole domain of photovoltaics, electric fence charging is a potentially significant new market for micro turbines. Electric fences are much more widespread elsewhere in the world than in the United States. Denmark, for example, has been using electric fences to enclose fields since before World War II. Electric fencing is also quite common for managing sheep in New Zealand. A 12-volt micro turbine and a weatherproof battery-pack are all that's needed.

ELECTRIC VEHICLE CHARGING

Charging electric vehicles with wind energy is not yet an important market, but the potential remains. When not needed, excess generation from a household-size turbine could be dumped into the home's electric car.

TELECOMMUNICATIONS

One of the early applications for small wind turbines, as well as photovoltaics, was powering remote telecommunications sites. These were often located on inaccessible mountaintops, where supplying diesel fuel was difficult and expensive. The proliferation of new

Village power system. *The original twelve Vergnet 12-kilowatt turbines in this wind-diesel system on La Désirade met 70 percent of the island's electricity supply. Since their installation in 1992, the turbines have been upgraded to 25-kilowatt and another eight units added. Now the island is a net exporter of electricity to nearby Guadeloupe in the French West Indies (Vergnet SA).*

telecommunications companies and the explosion in the use of cellular phones could require installation of more wind-diesel hybrids. Like off-the-grid systems for remote homes, these hybrids also include solar panels and batteries.

VILLAGE POWER

According to Trudy Forsyth at the National Renewable Energy Laboratory (NREL) in Colorado, 50 percent of rural dwellers worldwide—about two billion people—don't have access to electricity. This is potentially a huge market for wind and solar hybrids. Government agencies and most small wind turbine manufacturers are designing packages to provide power for Third World villages. The lessons they learn from these projects will be useful for off-the-grid systems in the developed world as well.

For example, NREL found that the combined battery and inverter efficiency in a wind and solar hybrid system for the Costa de Cocos resort in the Mexican state of Quintana Roo was only 50 percent. This is the kind of practical information needed by all those living off the grid.

Often there's little difference between a village power system in the Third World and an off-the-grid power system in the developed world. The components are the same. The difference is principally one of expectations. In Indonesia, twenty-five million people live beyond the utility network. A kilowatt of electricity goes much further there than in the United States or in Europe.

As part of an aid project in Indonesia, World Power and NRG Systems assembled fifty Whisper 600s with NRG's 45-foot (14 m) tower. The off-the-grid package was designed to provide 100 kWh per month at sites with a 12 mph (5.5 m/s) average annual wind speed. In the Indonesian setting, this is enough to power a village with several lights, a refrigerator, and a radio or TV.

In contrast, ENEL, the Italian national utility, installed five of Vergnet's GEV 5.5 turbines for isolated single-family homes not connected to its grid. Similarly, Vergnet has installed forty similar turbines at off-the-grid homes in metropolitan France. Each of these Vergnet turbines will easily produce five times the output of the Whisper 600s. While the Vergnet turbines may serve the needs of only one home in the developed world, they are veritable power plants in the developing world.

Cathodic protection. *Rafael Oliva's Bergey 1500 used for cathodic protection of the natural gas well in the foreground. This windswept site on the Patagonian steppes near Rio Gallegos, Argentina, requires extra stiff blades.*

Interconnected or Utility Intertie

The once promising idea of tens of thousands of small wind turbines whirring above farms, ranches, and homes across the United States, feeding electricity into the utility network, has never fully materialized. Yes, technically and legally it can be done. And there are thousands of small wind turbines distributed across North America doing just that—but not in the numbers first envisioned.

Following the oil crises of the 1970s, many consumers worried that utility prices would continue to rise for decades to come. This fear provoked keen interest in using small wind turbines to offset consumption of electricity from local utilities. These wind turbines would be interconnected, or intertied, with the utility on the customer's side of the kilowatt-hour meter.

Electric utilities, of course, had no interest in furthering such ventures, and placed numerous road blocks in the path of anyone who tried. But when the passage of PURPA (the Public Utility Regulatory Policies Act) in 1978

removed at least the legal obstacles to interconnection, small wind technology seemed poised for explosive growth.

In part the effort succeeded. There were more than 4,000 small wind turbines installed in North America for this purpose. Similarly, California's giant wind farms are also a direct result of PURPA and Congress' generous subsidies. But a host of problems beset wind turbines large and small. Some problems were technical: the turbines didn't work as well as expected. Some problems were commercial: firms entered and left the business so fast that it was hard to tell who was a manufacturer and who wasn't. And then the price of oil collapsed, and with it concern about rising utility prices. It was old pedestrian uses, such as battery charging and water pumping, that saved small turbine manufacturers from extinction.

The situation today is just the opposite of that in the early 1980s. Few small turbines are being installed for utility interties. Only committed environmentalists and the utilities

Small utility-intertie wind turbine. *During the early 1980s hundreds of small wind turbines were interconnected with electric utilities in the United States. This Enertech drives an 1,800-watt induction generator. It stands atop a tapered tubular tower at a home in the San Gorgonio Pass, near Palm Springs, California.*

themselves are interested in them, says Mike Bergey of Bergey Windpower. Unlike those living off the grid—who are dependent on their power systems and thus place a high value on the power their systems produce—utility customers are typically more cost-conscious.

In strict economic terms, small wind turbines seldom make sense where utility power is already available, and excess generation can't be banked with the utility through net-metering regulations. Under such conditions, the wind system has a long payback. Of course, it's unfair to demand that wind turbines meet artificial economic criteria when few consumers think twice about whether a houseboat, vacation cabin, or a pair of snow-mobiles parked on a trailer in the backyard are economic or not. Where states permit the "net metering" discussed later in this chapter, the economics are more attractive, though still daunting.

Surprisingly, it's North America's electric utilities—especially utilities in rural areas—that are the most interested in connecting small wind turbines to the grid. Now that an extensive network of lines exists, utilities have found it's expensive to maintain them. Utilities on America's Great Plains, for example, are learning that small wind turbines may offset some of the cost of upgrading or maintaining rural lines.

"We hope the domestic market [for inter-connected systems] will return some day," says Mike Bergey. It very well could. The American political climate could change overnight. Another war in the Middle East, a disastrous change in global climate, a Western reactor exploding in a populous region, or any one of a host of reasons could do it.

INTERCONNECTION TECHNOLOGY

Another difference between the early 1980s and today is the kind of small wind turbines being used for utility interties. In the 1980s most small turbines, such as those built by long-defunct Enertech, used induction generators. Today, nearly all small turbines destined for utility intertie use permanent-magnet alternators and inverters.

Induction or Asynchronous Generators

Wind turbines using induction generators are the simplest to connect to the grid. Early pro-motions by Enertech showed a wind turbine with an electrical cord for plugging into a wall outlet. Most of the medium-size wind tur-bines operating in wind power plants around the world are nothing more than glorified in-duction generators with an electrical cord. No mysterious black box is needed to convert the generator's output to the form used by the utility.

The induction generator, or asynchronous generator as it's known technically, uses cur-rent in the utility's lines to magnetize its field. Thus, the voltage and current the induction generator produces are always synchronized with that of the utility. In principle, the in-duction generator is unable to operate with-out the utility. When the utility's system is down, the induction generator is down too.

Electronics wizards can fool induction generators by using capacitors to charge the field, allowing induction generators to be used in stand-alone power systems. Vergnet wind turbines, for example, drive induction generators in this manner for use in wind-diesel and battery-charging systems in France and its overseas territories.

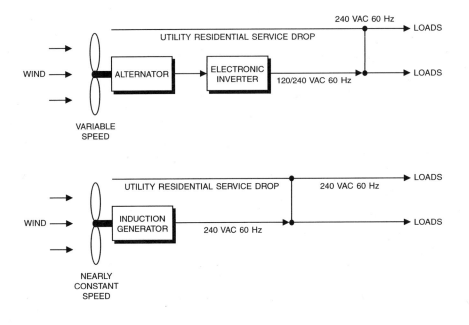

Utility-compatible wind systems. *Most wind turbines designed for utility intertie use either an induction generator or a permanent-magnet alternator and inverter. Modern solid-state electronics have greatly simplified the connection of small wind turbines to the utility network. Unfortunately, many regulatory and economic roadblocks remain in North America.*

Induction wind turbines do use sophisticated controls to tell the turbine when to connect and disconnect from the grid. If they didn't, the wind turbine would "motor" in low winds, acting like a giant fan and consuming electricity instead of generating it.

Inverters

It seemed miraculous when Windworks announced in the mid-1970s that you could use a Gemini synchronous inverter to connect a 1930s-era Jacobs wind turbine with an electric utility. Although the technology for doing so had been around for some time, it wasn't widely known among alternative energy advocates. The technology is much less mysterious today. The electronics are now quite commonplace, if not passé.

This inverter and the others that quickly followed took DC, in the case of the old Ja-

cobs generator, or rectified the variable-voltage, variable frequency AC from permanent-magnet alternators to DC, inverted it to AC, and synchronized it with the AC from the electric utility.

In this way, old synchronous inverters were line-synchronized, or line-commutated. They used SCR (silicon controlled rectifier) switches with analog controls to signal when they would feed bits of current into the utility system. Because they were line-commutated they needed the utility's line present to function. In the North American market, only Bergey Windpower and Wind Turbine Industries still produce turbines with line-commutated inverters.

Modern inverters are self-commutated. They can produce utility-compatible electricity using their own internal circuitry with IGBTs (integrated-gate, bipolar transistors) and digital controls. The new self-commutated inverters greatly improve reliability and power quality over the older line- commutated versions.

These self-commutated inverters use exactly the same technology as sine-wave inverters for off-the-grid power systems. These inverters use the DC from a battery storage system and produce an AC sine wave similar to, though not identical to, that of the utility. Thus, it's not surprising that modern sine wave inverters, such as Trace's SW series, offer the ability to feed excess power to the utility system.

With these new inverters, you can take any battery-charging wind turbine from Bergey Windpower, World Power Technologies, Southwest Windpower, or others and produce utility-compatible electricity. Unlike the old line-commutated systems, the new utility interactive systems require batteries to operate. In fact, they're simply stand-alone power systems connected to the utility through the inverter.

In these utility interactive systems, when electrical demand exceeds supply and the batteries are nearly exhausted, the inverter automatically draws power from the utility until the batteries are recharged. But when there's a surplus of generation relative to the load and the batteries are fully charged, the inverter can also feed the excess power back to the utility. Best of all, should utility power fail, in a storm for example, the inverter and batteries provide an uninterruptible power supply. The inverter automatically switches to a conventional, stand-alone battery system. You never notice the transfer from utility power to power from your own renewable resources.

Advances in inexpensive inverters for

photovoltaic panels may provide spin-off benefits to micro turbines. Some solar panels include their own 100-watt inverter, enabling you to literally plug the panel into a wall outlet. (Of course, only with a twenty-page contract from your local utility.) Manufacturers of micro turbines are investigating whether these inverters can be adapted to small wind turbines. If so, future micro turbines could begin to appear like contemporary computer peripherals: literally plug and play.

> Ideally, you would like to run your kilowatt-hour meter backward, selling any excess energy at the retail rate and buying what you need when you need it. In this way, you use the utility as a battery: The utility stores your energy until you need it.

DEGREE OF SELF-USE

Output from a wind turbine varies with the wind; electrical consumption in your home varies with the time of day as well. When you put the two together in a utility intertie system, you get an almost unpredictable mix: Some moments there will be excess generation; other times, there's a net deficit and you'll need to draw power from the grid.

Ideally, you would like to run your kilowatt-hour meter backward, selling any excess energy at the retail rate and buying what you need when you need it. In this way, you use the utility as a battery: The utility stores your energy until you need it. In North America, most utilities pay significantly less for the

Percent of Use on Site.

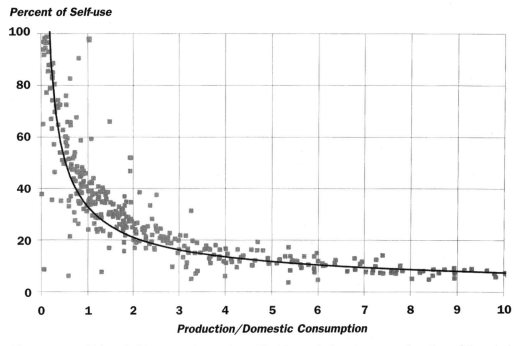

The percent of household energy demand supplied by a wind system as a function of the wind turbine's annual energy output and total household consumption. The most detailed records on this relationship come from Germany. This chart is derived from data on more than four hundred household-size wind turbines. There is some scatter in the data because some owners wisely shift discretionary consumption to periods when wind is available (Institut für Solare Energieversorgungstechnik, ISET).

energy they buy back than for the energy they sell. In practice, the utility will replace your kilowatt-hour meter with two ratcheted meters: one to register your purchases, the other to register your sales to the utility.

Homeowners with a wind turbine under these conditions want to minimize their excess generation. Why sell it to the utility for $0.03 per kWh when the utility will turn right around and sell it back to you for $0.10 per kWh? To avoid selling a surplus, homeowners have several choices: They can adjust their consumption by using dump loads as much as possible; they can match their consumption to wind availability as much as possible;

or they can use wind turbines smaller than they might otherwise select.

It's a Catch-22 situation: In utility interties, wind turbine cost-effectiveness increases with size; but to minimize the amount of energy you sell to the utility at a deep discount during times of surplus, you'll want to buy a wind turbine that produces only a portion of your domestic consumption. In Germany, for example, only one-third of the electricity from a wind turbine sized to meet the annual domestic needs of a home will actually be used in the home. Two-thirds will be sold back to the utility. To use two-thirds of the consumption in the home and only sell one-third back to the utility, you need to use a wind turbine that produces half or less of your annual domestic consumption.

NET METERING

To avoid this situation, twenty-two states in the United States permit net metering for wind turbines. In principle, net metering allows you to run your kilowatt-hour meter backward. Again, two ratcheted meters will probably be used so the utility can keep track of the inflow and outflow. The utility balances the account, usually every month.

Most states limit net-metering interconnections to 10 kW, though in some states the limit is higher: Minnesota, 40 kW; Massachusetts, 30 kW; New Mexico and North Dakota, 100 kW. There is no limit in Iowa. Other states permit net metering, but discriminate against wind turbines.

In most states, net-metering regulations affect only investor-owned or regulated utilities, thus excluding the many consumers connected to rural electric cooperatives. Only eleven states offer net metering on both investor-owned utilities and rural co-ops. (To find out if your state permits net metering, contact the American Wind Energy Association.)

Net metering makes interconnecting a small wind turbine with the utility far more financially attractive than it would be otherwise, and every state should have at least net-metering provisions. Even so, at the cost of today's small wind turbines, and the low prices for electricity in much of North America, it's a tough sell. You must want to do it for reasons other than economic factors alone. Remember that many of the great public works projects in North America, such as

dams and canals, make no economic sense whatsoever, but we as a society have forged ahead anyway.

POWER QUALITY AND THE UTILITY

Always consult the local utility before interconnecting your wind turbine with their lines. The utility may have some valid concerns about the power factor, voltage flicker, and harmonics produced by your wind machine. They may also have concerns about the safety of their personnel when working in the neighborhood. These concerns are no reason to deny a homeowner an intertie, but the utility has a right to ask these questions and to receive the answers before they accept an intertie with their lines. The utility may also require payment for reasonable costs arising from the interconnection. This shouldn't be much, but get it spelled out beforehand. Of course, it may take an attorney and a big bank account to convince the utility to do what they're obligated to do.

We now have more than two decades of experience with interconnected wind turbines. By the start of the new millennium, wind turbines will have operated more than three billion hours on the lines of electric utilities in Europe, the Americas, and Asia without creating the electrical problems utilities once feared.

EUROPEAN DISTRIBUTED GENERATION

The small turbine industry in Denmark began much the same way as it did in North America. The first turbine using an induction generator was connected to the grid without

Household-size wind turbine. *The European boom in wind energy has rekindled interest in small utility-intertie wind turbines. The 6-meter (20-foot) diameter downwind rotor on this 5-kilowatt Kolibri drives an induction generator mounted inside the tower. The multiblade fan tail orients the rotor downwind of the tower.*

permission and unbeknownst to the utility in the mid-1970s. But the Danish industry has since followed a far different path than the small turbine industry has in the United States. The once small turbines of Danish manufacturers have gradually grown larger, becoming the core of the world's commercial wind power industry.

Unlike in the United States, where many of the early utility intertie turbines have fallen into disrepair and stand forlorn and inoperative, there are two hundred household-size turbines from the 1970s and early 1980s still operating in Denmark. More than 80 percent use induction generators and are still interconnected with the utility. They represent 2.6 megawatts of capacity, about 5 percent of that installed in Denmark in 1998.

These turbines include the stodgy, technological forebears of today's medium-size Danish machines, as well as more novel designs. For example, nearly half of these early turbines are Kuriants, an unusual design for Denmark because their 12-meter (40-foot) three-blade rotor is downwind of the tower. Most Danish turbines are upwind designs. Kuriants also use guyed towers, a feature common in the United States but rarely seen in Europe.

Early Danish turbines were about the same size as those being installed in the United States: 7 kW, 10 kW, and 15 kW. But the turbines and their manufacturers grew because there was political support for them to do so.

In Denmark—and since 1991 in Germany as well—wind turbine owners are paid a fair price for the electricity they produce. Germany's Electricity Feed Law, the *Stromeinspeisungsgesetz,* requires utilities to buy wind-generated electricity at 90 percent of the retail rate. Danish utilities are required to

pay 85 percent of the retail rate. In addition, the Danish government refunds a carbon dioxide tax collected on all electricity generated in the country. Wind turbines in these two countries earn about $0.10 per kWh. This contrasts sharply with the $0.02–$0.04 per kWh paid in North America in states without net metering.

By the start of the new millennium, wind turbines will have operated more than three billion hours on the lines of electric utilities in Europe, the Americas, and Asia without creating the electrical problems utilities once feared.

In Denmark and Germany, it's in everyone's interest to install the size wind turbine that makes the most economic sense. People are not forced to choose a wind turbine smaller than they need because the buyback rate is so low.

The result has been the steady growth of distributed wind turbines in Germany and Denmark. About two-thirds of the several thousand wind turbines installed in the two countries are owned by farmers or cooperatives of local residents, a reality not unlike the image once envisioned by proponents of wind energy in North America.

If there were Renewable Energy Feed-in Tariffs (REFITS) in the United States and Canada such as the electricity feed laws in northern Europe, we would quickly find interconnected wind turbines blossoming again in North America.

Hybrid wind and solar. *This Proven WT600 works in conjunction with the photovoltaic modules to power a remote telecommunications site in Scotland (Proven Engineering).*

6

Siting and Safety

One of the most challenging aspects of using wind energy is finding a place to put the wind turbine and tower. If it's too close to your home, the wind turbine will suffer from the building's interference with the wind. (You may also suffer complaints from your family about the noise, especially if you mount the turbine on your house.) If it's too far away, the cost of cabling and burial rises prohibitively, as do losses from electrical resistance. Every installation is a balance of these factors. Rarely is there an ideal site.

Another often overlooked aspect is safety. Though small, these machines are not toys. They can kill or maim—and have!

TOWER HEIGHT

Jason Edworthy's experience at Nor'wester Energy Systems in Canada convinces him that the old 30-foot (10-meter) rule of thumb is sacrosanct. This classic rule from the 1930s dictates that for best performance, your wind turbine should be at least 30 feet (10 meters) above any obstruction within 300 feet (100 meters) of the tower. Under the best conditions, a tower height of 30 feet (10 meters) is the absolute minimum, says Edworthy.

The reason is clear: The tower needs to be tall enough to clear the zone of disturbed air flow around buildings, trees, and other obstructions. This zone of turbulence can extend up to twenty times the height of the obstruction downwind, and surprisingly, up to twice the height of the obstruction upwind. Near the obstruction—your house, for example—the disturbed zone can reach twice the obstruction's height. If your house is 30 feet (10 meters) tall and you plan to install the tower nearby, the tower will need to be at least 60 feet (20 meters) tall. In this case, the 30-foot (10-meter) rule works also. The tower should be 30 feet (10 meters) above the house, or 60 feet (20 meters) tall.

For Inuit outposts in Canada's Northwest Territories, Edworthy designed an easily transportable package of turbine and tower. The Inuit fur trappers installed their Windseekers on 40-foot (12-meter) guyed masts to

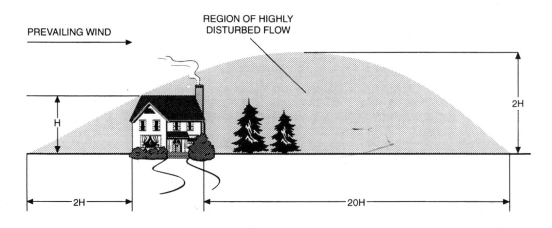

PREVAILING WIND

REGION OF HIGHLY
DISTURBED FLOW

H

2H

2H

2H

20H

Turbulence and tower height. *Power-robbing turbulence is created by trees, shrubs, buildings, and other obstructions. Ideally, you want to install your wind turbine in a zone of undisturbed flow, just as you want your solar panels in full sunlight.*

feed four 100-amp-hour batteries in a simple 12-volt system. Edworthy insisted on the 40-foot (12-meter) towers as the minimum acceptable tower height, even for these extremely remote sites where transport was difficult and costly.

In Denmark, manufacturers of household-size turbines supply their machines with 12-meter (40-foot) or 18-meter (60-foot) towers. They don't offer towers less than 12 meters tall.

Until recently, it was difficult to meet such minimum height requirements at low cost. Often a Rohn 25G guyed tower was the least costly option in North America. Rohn guyed towers use lattice tower sections for the mast. The Rohn 25G was the lightest weight mast then available. The only other alternative was free-standing lattice towers that were even more expensive. (For more information on how the Rohn 25G tower can be used with household-size turbines, such as the Bergey 1500, see *Wind Power for Home & Business.*)

Fortunately, there has been significant progress in the development of low-cost guyed towers for micro and mini turbines. World Power, Southwest Windpower, and Lake Michigan Wind & Sun provide kits for guyed, tilt-up towers. The homeowner provides the steel pipe needed for the mast.

Even more promising has been NRG Systems' adaptation of their lightweight anemometer masts for use with small wind turbines. The NRG masts have become the defacto standard among professional meteorologists and wind prospectors around the world as an inexpensive and easily erectable mast system.

NRG's guyed towers use thin-walled steel tubing. This tubing is not as thick as steel pipe and must be carefully matched to the wind turbine it will be supporting. NRG offers masts in three diameters: 3.5-inch (89 mm), 4.5-inch (114 mm), and 6-inch (152 mm). The 3.5-inch (89 mm) mast is suitable for some micro turbines. The 4.5-inch (114 mm) mast is designed for Bergey Windpower's 850. The 6-inch (152 mm) tubing will support a Bergey 1500.

NRG's towers are available in four sizes: 44 feet (13.4 meters), 64 feet (19.5 meters), 84 feet (25.6 meters), and 104 feet (30.7 meters). NRG provides the towers as a complete system, including base plate, gin pole, screw anchors, and guy cable. The guy cable comes sized with cable clamps already swaged in place. The tower sections slip together, simplifying assembly.

> To cut corners, some homeowners have mounted their micro turbines on the roof. Don't bother. It's not worth the trouble.

With the advent of inexpensive tower kits and the introduction of NRG's tilt-up guyed mast systems for small wind turbines, there have never been greater opportunities for installing towers of the proper height.

Mick Sagrillo agrees with Edworthy about the importance of using tall towers. To emphasize this, Sagrillo says, "If it's less than 30 feet (or 10 meters), don't bother." He recommends tower heights of at least 60 feet (19 meters) for micro turbines, 80 feet (24 meters) for turbines up to 1,500 watts, 100 feet (30 meters) for 10 kW machines, and 120 feet (37 meters) for 20 kW units and larger.

Shorter towers may be acceptable in some conditions: In electric fence charging, an application where micro turbines compete directly with photovoltaics and one where the turbine may be moved frequently, a shorter tower makes sense.

At sites in Scotland, Proven turbines are installed on free-standing towers only 6.5 meters (20 feet) tall. Hugh Piggott finds that in open terrain these work acceptably. Equally

important, these short towers are easier to move through the planning approval process in Britain than are taller towers.

Southwest Windpower's Andy Kruse contends that what tower height people use with Southwest's Air 303 depends on what they want from the micro turbine and what they are willing to accept. Kruse argues that the Air 303 is so inexpensive that installing it on a taller and more expensive tower is seldom justified. "Sure, if the wind turbine costs several thousand dollars, it makes sense to put it on a taller tower," he says. The average tower height for the Air 303 is only 25 feet (7.6 meters), according to Kruse. "People feel comfortable with this tower height. Admittedly, this is not the optimum, but users shouldn't be so concerned with tower height if they're willing to accept less than optimum performance."

> With the advent of inexpensive tower kits and the introduction of NRG's tilt-up guyed mast systems for small wind turbines, there have never been greater opportunities for installing towers of the proper height.

To cut corners, some homeowners have mounted their micro turbines on the roof. Don't bother. It's not worth the trouble.

Others have mounted their wind turbines on tops of trees. This is never a good idea. It spells trouble for the wind turbine because of turbulence, trouble for the user because of

Building-stayed tower. *While the tower here is not ideal, it is far superior to roof mounting. Using the building to stay the tower may eliminate the need for guy cables. Note that the tower is attached to an unoccupied building. The turbine's noise and vibrations will not annoy anyone here. This Windseeker 502 is part of a wind-solar hybrid for an off-the-grid home near Tehachapi, California. In fact, all the homes in this subdivision are off the grid and many use wind-solar hybrids.*

the need to climb the tree to service the machine, trouble for the tree because the turbine will eventually kill it, and trouble for wind energy in general because the public will tend to question whether wind energy is really worth that much fuss.

Sometimes wind energy isn't the right choice. If you live in a forest of tall trees and you can't afford a tower tall enough to clear the trees, then wind energy isn't for you.

TOWER PLACEMENT

Ideally, the turbine and tower should be well clear of any buildings and obstructions. If there's a hill nearby, place the turbine on the hill, even if it means a longer cable run.

It's also important that the anchors and guy cables for guyed towers be well beyond traveled ways—roads, vehicle tracks, and footpaths. If there are any farm animals roaming the site, guyed towers and their anchors must be fenced or otherwise protected.

There have been some cases where a micro turbine has been installed on a tower adjoining an unoccupied building. This is not an ideal solution, but it can work.

NOISE

Most small wind turbines are noisy when governing in high winds. There's no escaping that fact. Some are worse than others. The Air 303, because its blades flutter in strong winds, is particularly noisy. And Whisper wind turbines are anything but a whisper when governing.

Small wind turbines are as noisy quantitatively as medium-size wind turbines that produce 100–1,000 times the power. Whether you find the noise bothersome or not depends upon the wind speed, how close you are to the turbine, and whether it's your machine or not.

Noise from small wind turbines is most noticeable in high winds. Some have characterized the sound from governing micro turbines as being like that of a chain saw. But even in light winds a small wind turbine may sound like a flock of starlings.

The simplest way to reduce noise is to move the turbine as far away as is practicable. This often is also the best solution for minimizing the turbulence affecting your turbine and for improving energy capture—a win-win situation.

Though you may find the noise distracting at first, you could just as easily grow fond of it over time. The realization that the turbine is working hard on your behalf can gradually persuade you that what once seemed a whirring noise now sounds like a cat's soft purr. Your neighbors, however, may not be quite as tolerant.

URBAN WIND

Just as there's no typical rural environment, there's no typical urban environment either. It's safe to say that in the heart of urban centers, wind energy just doesn't make sense. Sure, some activists installed a turbine atop a tenement in the Bronx two decades ago and there are Bergeys installed atop high-rise buildings in Washington DC, Dublin, and Melbourne, but none are operating. For practical purposes, it doesn't work.

There are open spaces in every urban center where small wind turbines may be appropriate: Parks, preserves, and athletic fields in or near major cities all offer potential for demonstrating how small wind turbines work. For example, there's a Bergey at an Audubon

Well-exposed tower. *This Vergnet SA turbine in the Corbières of southern France is a good example of locating the turbine as far from obstructions as is practicable. The tilt-up tubular tower also elevates the wind turbine well above surrounding obstructions (Vergnet S.A.).*

center just outside Pittsburgh and another at a park in the Sacramento River delta.

Suburbia provides more opportunities than built-up urban centers, but tracts in subdivisions seldom permit such "nonconforming" land uses as installation of a wind turbine. A simple rule of thumb is that zoning boards will usually allow structures if there's enough room for it to fall over without affecting neighboring property.

SAFETY

Of the nearly twenty people killed working around wind turbines worldwide in the past twenty years, two have been killed working on small wind turbines. Both fell off the tower. There have been numerous close calls. Here are a few suggestions (for more details see *Wind Power for Home & Business* and Hugh Piggott's *Windpower Workshop*):

Moving Machinery

Wind turbines contain rotating machinery, and every warning about how to work around such machines applies. Don't allow long hair, loose clothing, rings, or necklaces around any turning shaft—regardless of how slowly it's turning. Hugh Piggott, in *Windpower Workshop*, recounts a "hair raising" tale told by Mick Sagrillo of the encounter between Mick's pony tail and the slowly turning shaft of a "Jake" in an Alaskan shop. If you have long hair at least keep it tucked into your shirt or under your hat.

Never go near a spinning wind turbine. If you have to work near the turbine, furl it and brake the rotor to a stop. If the wind is light to moderate, you can bring most rotors driving permanent-magnet generators to a halt by shorting all phases in the armature. World Power includes just such a brake switch with all its new control panels. You can order such a switch for the Air 303 from Southwest Windpower. You'll have to build your own for the Bergeys.

Electrical

Permanent-magnet alternators produce a voltage whenever the rotor turns—even when disconnected from the load or control panel! Before servicing the control panel, disconnect the power supply from the turbine. (Install a fused disconnect switch for this purpose. It will come in handy.) Even though you may be using a low-voltage DC power system, many wind turbines produce three-phase AC and when unloaded can reach very high voltages.

Off-the-grid power systems can experience high current draws and high charging rates. Both conditions require that all cabling be amply sized and the connections terminated correctly, for safe operation.

Fuse all power sources (both wind and solar, for example), both AC and DC loads, and connections to the batteries. Trace, Pulse Energy Systems, and others supply pre-engineered, assembled panels that include built-in fuses or circuit breakers for both the DC and AC side of off-the-grid power systems. These are available under several trade names. Some are approved by various standards or rating organizations, such as Underwriters' Laboratories, and will pass muster with building inspectors in much of the United States. They are part of an encouraging trend toward more standardized and professional DC-to-AC power systems. Use them.

If you have any doubts about how to properly fuse a part of your power system or how to make sound terminations, consult the manufacturer or supplier of the component.

Working atop a tower
is always dangerous.

Batteries

Always use extreme caution when working around batteries. Use the same precautions you would use when working near an automotive battery. Wear goggles to protect your eyes from a spray of battery acid, should an accident occur.

Beware of dropping metal tools onto exposed battery terminals. This is a recipe for disaster. Some pros recommend insulating metal tools for working around batteries.

Vent the batteries adequately to the outdoors to prevent concentrations of explosive hydrogen gas. Avoid having any source of sparks or open flame around the batteries.

As when working around rotating machinery, don't wear rings or necklaces around batteries.

Towers

Working on or around wind turbines and their towers poses two kinds of hazards. One is a fall from the tower; the other is being hit by something falling from the tower.

Tower safety. *Remember, you have to service it, too! Most small wind turbines have no provision for a work platform. Gaia's prototype 5-kilowatt turbine, like most household-size wind turbines in Europe, includes a work platform. Here Hallgrimur Halldorsson is troubleshooting the monitoring instruments on Gaia's prototype at the Folkecenter for Renewable Energy's test field. Note Halldorsson's use of the safety harness and lanyard, even while on the work platform.*

As Hugh Piggott often notes, "bits and pieces" fall off the thirty-odd turbines he maintains near his home in northwestern Scotland. So never work beneath an operating wind turbine. Turn it off or brake it to a stop before performing any service at the base of the tower. Even a small nut can pick up a lot of speed from 60 feet (18 meters) overhead

and cause a nasty injury if anyone is unlucky enough to be in its way.

Ideally, guyed towers should be located two tower lengths away from occupied buildings and from power lines. The reason for this may not be apparent. While it's clear that you or someone else could be injured by the mast if it fell over, what's not so obvious is that the guy cables can go slicing through the air well beyond the end of the mast. The whipping guy cable could not only cause serious injury, but also could lash a bare conductor on a nearby utility line, making the whole tower and all its guy cables electrically charged.

Servicing

All wind turbines must be regularly inspected and serviced. Sagrillo likes to quote Jim Sencenbaugh on the need for routine inspection and service: "The life of a wind turbine is directly related to the owner's involvement."

It's important to anticipate how you will get to the wind turbine. Working atop a tower is always dangerous. "Avoid it if at all possible," says Scoraig Wind Electric's Piggott. If you're using a free-standing lattice tower or a heavy-duty guyed mast, climbing may be your only option. (Homeowners seldom have access to the bucket trucks or light-duty cranes that can make servicing a wind turbine both simpler and safer.)

Always wear and use an approved safety belt and lanyard when climbing and working atop a tower. (*Home Power* magazine occasionally carries articles on how to use basic safety equipment. Belts and lanyards are also discussed in *Wind Power for Home & Business.*)

Climbable towers should always include a fall-arresting or safety cable. When ascending or descending the tower, attach your safety belt to the cable with the special sliding shoe that's provided. (Towers on all wind farm turbines use this system.)

A work platform should always be included on climbable towers. This need not be elaborate, but should be sufficient to allow safe and comfortable servicing of the wind turbine. Be advised that most towers made in the United States for small wind turbines don't include work platforms!

Tilt-up, guyed tubular masts must be lowered to allow servicing of the wind turbine. This entails its own set of hazards, but it does eliminate the need, at least for small wind turbines, for working on the turbine in the air.

Lattice masts on tilt-up, guyed towers enable you to climb the tower for minor service. Bergey Windpower recommends using guyed lattice masts on its bigger turbines for this reason. Again, if the mast is intended to be climbed, it should include a work platform and fall restraint system.

Raising and lowering tilt-up towers is also risky because of the heavy loads on the guy cables when the tower is near the ground. Never stand underneath the tower or guy cables when the turbine is being raised or lowered. Something unexpected can go wrong. In the dairy state of Wisconsin, windsmiths call guy cables on tilt-up towers "cheese slicers" because of their potential to catch someone unaware.

An experienced field crew at the Alternative Energy Institute at West Texas A&M was once lowering a 10-meter diameter, 25 kW turbine at their test field. They had done this many times before. But this time was different. There was a miscommunication. And to make matters worse, there was a photographer in the path of the tower. The tower whizzed inches by the photographer's head and crashed to the ground. No one was hurt but there were a lot of deep breaths and red faces. The photographer quickly left Texas, never to return.

Rustic work platform. *This old Aerowatt provides reliable power to a Chilean ranger station protecting the nesting sites of Patagonia's rare Magellanic penguins (Nancy Nies).*

7

Buying Small Turbines

Wind turbines, large or small, are not cheap. Nor for that matter, are solar modules, batteries, inverters, and the other components needed in a home power system. Many homeowners have no idea how much electricity they consume, and mistakenly think a micro or mini turbine will power their entire home.

"Don't expect to spend $500 and become energy independent," cautions Nor'wester Energy Systems' Jason Edworthy. "You have to spend enough money to do it right," he says, and that's usually a lot more money than many expect.

The best place to begin evaluating the price of a small wind system is to realistically determine what it is you want. Do you want a turbine to meet the limited needs of a vacation cabin used only on weekends? Or do you want the turbine as a complement to the photovoltaic panels and the stand-alone power system you already have? Sum the anticipated electrical consumption you want the turbine to meet, estimate the wind available, then determine the size turbine you need.

Next, compare the various wind turbines in the size class you need. This is quite subjective because you must weigh not only price but less tangible factors such as quality and reliability. Avoid buying any product based on price alone; instead, look at relative price to determine what's a better buy.

One common—though unreliable—measure is the price of the turbine per kilowatt ($/kW). Since there's no standard rating system for small wind turbines, products with high power ratings have the lowest price per kilowatt, but they may not produce as much electricity as a turbine with a more realistic power rating. A better measure is the price of the turbine relative to the area swept by its rotor ($/m², or $/ft²). This measure is helpful because it's not subject to the vagaries of power rating.

The National Renewable Energy Laboratory now uses what they call a "cost performance ratio" to evaluate the cost-effectiveness of small wind turbines, says NREL's Jim Green. The ratio is simply the purchase price of the wind turbine divided by its estimated

Small Wind Turbine Prices.

Relative Price ($/m²)

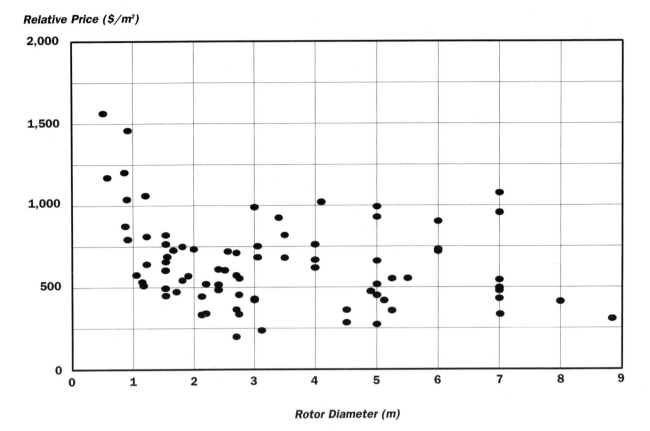

Rotor Diameter (m)

The relative cost of small wind turbines decreases with increasing size. Though the total price increases with size, the cost per kWh decreases.

annual energy output. This method eliminates any confusion over various power ratings. To use this method effectively, you need to know the wind resource and how much energy each turbine will produce under those conditions.

As a rule, wind energy becomes more cost-effective as the wind turbines increase in diameter. The relative price (price per kilowatt, price per swept area, or cost performance ratio) decreases with increasing rotor diameter. Though smaller wind turbines cost less, they are proportionally more expensive.

The wind turbine is only part of the system. Towers are essential too.

TOWERS

For micro turbines, a quality tower will cost as much as the turbine, sometimes more. The least costly and most user-friendly tower option is a tilt-up, guyed tubular mast.

Some homeowners have used wooden utility poles; others have used old farm windmill towers. In either case, the turbine must be adapted to fit the tower. This isn't as

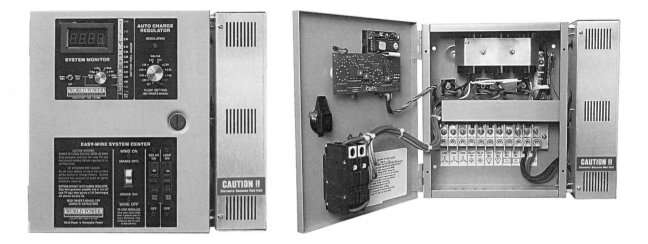

Control panel for battery-charging wind turbine. *World Power's system center provides clearly marked terminal blocks, ample controls—brake switch, charge equalization, DC circuit breakers—and a built-in dump load, all in one package. If a wind turbine doesn't come with these features, you will have to provide them (World Power Technologies).*

straightforward as it may seem. The clearance between the rotor blades and the tower is especially critical. There must also be a way to install the turbine, and service it once installed.

To cut costs, some people try to build their own towers. "Don't do it," advises USDA's Nolan Clark. The tower should be matched to the turbine, and like the rotor, it's a critical component. Several suppliers offer tower kits that are preferable to building a tower from scratch. These kits require enough home assembly to satisfy the most ardent do-it-yourselfer.

NRG's tower system for micro and mini turbines offers the best compromise between price, engineered reliability, and ease of installation. The NRG system relies on a floating mast and screw anchors. No soil excavation or placement of concrete is necessary. Each tower system includes the cables, anchors, gin pole, and hinged base plate

needed for raising and lowering the tower.

The NRG mast rests on a metal pad that's simply staked to the ground surface. The guy anchors are screwed into the ground by hand. (They can also be power driven or anchored by truck-mounted augers.) Where soils are unsuited for screw anchors, or where frost heave causes the mast to rise and fall with the seasons, the mast pad and screw anchors can be replaced by conventional concrete anchors.

CONTROL PANELS

A control panel isn't absolutely necessary for a micro or mini turbine. Some manufacturers, Southwest Windpower for example, have built their battery-charging controls into the generator. Many small turbines will be used in hybrid systems, where other electronic components often provide the supervisory functions of a controller. Still, a control panel is more than just convenient.

It's helpful to know how the wind turbine is performing, so that when it's not, you can get it fixed, and it's helpful to have a brake switch, so when you need to stop the turbine, you can. It's also useful to have over current protection, so that when something goes wrong, you reset the circuit breaker or replace the fuse and not the wind turbine. If a control panel with these functions isn't included with the wind turbine, the user will often have to provide it.

It's buyer beware in the used wind turbine market. If someone tries to sell you a supposedly "new" turbine at a discount price, think twice.

World Power offers an attractive control panel designed to accept power cables from both a wind turbine and from solar panels. The clearly marked terminal blocks also offer connections for direct DC loads, battery connections, and a built-in dump load. (The attached resistance air heater could easily be replaced with a water heater, for example.) World Power's system center, as they call it, includes a brake switch for short-circuiting the turbine's armature. In light winds, activating the switch will bring the rotor to a halt.

Compared to the nonexistent—or at best, skimpy—panels offered by some manufacturers, World Power's panel adds significant value to the wind system. And though price is important, value is more so.

PRICE

The prices in the accompanying table are based on an appropriately sized NRG tower and a set of batteries representative of each wind turbine size class. The micro turbine system, for example, is 12 volts; the mini turbine system is 24 volts; and the household-size system is 48 volts. These prices are only approximate and subject to change.

Prices in the table are about the minimum needed to do the job right. You can shave a few dollars here and there, but even a sharp buyer will find that the final outlay will be within 10 percent of the prices listed.

To ensure more successful applications of small wind turbines, manufacturers must begin selling complete hybrid systems and not merely a collection of components. "You must be able to buy a matched system," says USDA's Nolan Clark. Mail-order companies such as Real Goods, Jade Mountain, and Alternative Energy Engineering can provide helpful advice on matching components. Some vendors offer complete packaged systems.

World Power, for example, offers competitively priced packages that can include the turbine, tower, batteries, inverter, and wiring kit. They package their H900 in a 24-volt system with eight 220-amp-hour batteries and a 1500-watt Trace (DR 1524) inverter. You'll need a tower. You could do without the inverter and instead rely on DC appliances, or you could skimp on the batteries, but no matter how you cut it the final price will fall between $3,000 and $4,000—and this is for a minimalist system.

Typical 1998 Prices without Installation for Battery-Charging Wind Turbines (totals subject to rounding).

Micro Turbines

	Ampair 100 0.92 m (3 ft)	Marlec Rutland 913 0.91 (3 ft)	SWP Air 303 1.1 m (3.7 ft)
Turbine	$1,000	$600	$600
40-foot (13 m) tower	$800	$800	$800
Batteries (2 220-amp-hour)	$200	$200	$200
Misc. (20%)*	$400	$320	$320
	$2,400	$1,900	$1,900

Mini Turbines

	SWP Windseeker 503 1.5 m (5 ft)	WPT Whisper H900 2.1 m (7 ft)	BWC 850 2.4 m (8 ft)
Turbine	$1,100	$1,600	$2,200
60-foot (19 m) Tower	$1,000	$1,000	$1,200
Batteries (4 220-amp-hour)	$400	$400	$400
Misc. (20%)*	$500	$600	$760
	$3,000	$3,600	$4,600

Household-sized Turbines

	WPT Whisper H1500 2.7 m (9 ft)	BWC 1500 3.1 m (10 ft)	Proven WT2500 3.5 m (11 ft)
Turbine	$2,700	$5,000	$6,500
80-foot (25 m)	$2,500	$2,500	$2,500
Batteries (8 450-amp-hour)	$800	$800	$800
Inverter (Trace SW4024)	$3,100	$3,100	$3,100
Misc. (20%)*	$1,820	$2,280	$2,580
	$10,900	$13,700	$15,500

*Includes freight, switches, fuses, cable, junction boxes, enclosures, etc.
SWP = Southwest Windpower; WPT = World Power Technologies; BWC = Bergey Windpower

DO-IT-YOURSELF TURBINES

If you're an inveterate tinkerer and you're determined to build your own wind turbine, contact the Centre for Alternative Technology in Wales or Scoraig Wind Electric in Scotland. Both offer plans for do-it-yourself wind turbines. Kragten Design in the Netherlands sells a kit for building a mini turbine.

USED WIND TURBINES

There's a limited market for small used turbines. Check the classified ads in *Home Power* magazine. Unless you've seen the turbine firsthand, and know what you're looking at, it's best to buy only from reputable reconditioners, such as Lake Michigan Wind & Sun.

Be wary of any used equipment from

Do-it-yourself plans. *The Centre for Alternative Technology in Wales offers plans for a Cretan sail windmill for those determined to build their own wind turbine.*

Californian or Hawaiian wind plants. "They've worked pretty hard," says Wind Turbine Industries' Steve Turek, in classic Midwestern understatement. "It's like buying a used lawn mower from a golf course," he says. These machines have been beat to death by those who are pros at squeezing every last cent out of a piece of machinery. Buy a used wind farm turbine only if you've got a strong stomach, and an even stronger bank account.

It's buyer beware in the used wind turbine market. If someone tries to sell you a supposedly "new" turbine at a discount price, think twice. There are no "new" Enertechs, for example. The company went bankrupt a decade ago. If it's a dirt-cheap but "new" Jacobs Wind Energy Systems turbine, call Wind Turbine Industries and confirm that it has in fact been newly assembled. (Again, these are not 1930s-era "Jakes," though they use the same name.) Remember, if the deal you're offered is too good to be true, it probably is.

COLLECTIVE BUYING POWER

In a recent government aid program to the Third World, United States manufacturers sold their wind turbines at bargain-basement prices, while at the same time charging their North American buyers full fare. The government agency didn't accept the initial quotes and used its buying clout to force prices down. If homeowners and other individual buyers could organize multiple-turbine purchases, as the government agencies did in the aid program, they could possibly negotiate similiar deals.

Group purchases have yet to materialize, but they offer promise. USDA's Nolan Clark still sees hope for consumer consortiums that buy wind and solar hybrids collectively from manufacturers.

SUBSIDIES

Subsidies, or the more politically correct "incentives," are generally not available for small wind turbine projects in North America. As this book went to press in early 1999, federal energy tax credits in the United States were applicable only to commercial installations where wind-generated electricity was sold to a utility or other buyer. This precludes small wind turbines for home use.

Utility restructuring in California has created a pool of funds to subsidize experimental technologies, such as small wind turbines up to 10 kW that are connected to the grid. There are strings attached to the subsidy. The most significant requirement is that small wind turbine designs must be certified, or have at least one year of operation, in wind speeds of 12 mph (5.5 m/s) or better.

There are some minor incentive programs in several states and provinces. Contact the American and Canadian wind energy associations for a current list of North American subsidies by region.

Owner-installed wind turbine. *The author and his wife installed this wind turbine in California's Tehachapi Mountains without using a crane, electric winch, or truck.*

8

Installing Small Turbines

For micro and mini wind turbines, a tilt-up, guyed tower is both cheaper and easier to install than other tower choices. While installing such a tower isn't a snap, it's much simpler than erecting a free-standing lattice tower with a crane. There's also the advantage of being able to raise and lower the hinged tower whenever you want to service the turbine.

Students in Mick Sagrillo's class for Solar Energy International in Colorado installed a mini turbine on an NRG tilt-up tower within four hours, reports Paul Wilkins of *PV Network News*. Of course, Sagrillo knows what he's doing. He's installed or repaired seven hundred small turbines himself. Novices will require much more time to do the same job safely.

Tilt-up towers such as NRG's, use a gin pole to raise the mast. The gin pole is effectively a second, shorter mast at a right angle to the tower that acts as a lever to reduce the lifting loads.

NRG's 3.5-inch (89 mm) tilt-up mast system is shipped in 7-foot (2.1-meter) lengths. This allows transport by parcel delivery ser-

vices worldwide. Bundles of three sections in this size are easy to maneuver. But the 4.5-inch (114 mm) and 6-inch (152 mm) sections are shipped in 10-foot (3-meter) lengths, in bundles of three pieces, and require transport by motor freight. These cartons are hefty and require some effort to move around, but the individual sections are easily manageable.

TOWER RAISING

Everyone who has chosen a tilt-up tower to support a small wind turbine has had to face the difficult question of how to raise it. Next to servicing a wind turbine atop a tower, there is no more dangerous aspect of using wind energy than raising and lowering a wind turbine and its tower.

The most common technique among do-it-yourselfers in the United States is to raise tilt-up towers with a truck or tractor. Unless you're skilled with hand signals and with working around heavy equipment, raising a tower with a truck or farm tractor can be an extremely ticklish operation.

Raising an NRG tower. *Students led by Mick Sagrillo raise a Bergey 850 on an NRG tower near Carbondale, Colorado (Mick Sagrillo).*

Properly using a vehicle for raising a tilt-up tower demands a large crew. Sagrillo recommends two on the truck (one to drive, and one to watch for hand signals) and one for each anchor, or a minimum of six people. Gathering such a large group together can present a challenge. When the inevitable glitches arise, it puts one in the awkward position of either asking everyone to come back another day or forging ahead and taking chances one shouldn't. You can quickly wear out your credit with friends and family if the tower raising doesn't go as planned. You don't want a bunch of your friends standing around asking, "Hey, are we going to install this windmill or not?"

Communal tower raising can be a reward-ing experience, like Amish barn-raising, bringing people together for a common purpose. But barns last indefinitely. You put it up, and it stays up. Not so with a wind turbine. Whether we like it or not, small wind turbines do need repairs and we have to bring them down before we can haul them off to the local repair shop. Some are up and down a lot. Gathering six people together every time you want to raise or lower your turbine can quickly become tiresome.

GRIPHOISTS

Fortunately, griphoists are a common alternative where cranes are either too expensive or too difficult to use. Hand-operated griphoists are used throughout Europe and Canada for a variety of applications that include raising wind turbines and meteorological masts.

"It's a good way to raise a windmill," says Scoraig Wind Electric's Hugh Piggott. It gives you "plenty of time to check things." Piggott uses a griphoist to install wind turbines in Scotland. Niels Ansø uses a griphoist to raise and lower small turbines at the Folkecenter for Renewable Energy's test field in Denmark.

To Piggott, this tool is a *tirfor*. To Ansø, it's a wire *talje* (hoist). Jim Salmon, a Canadian meteorologist, calls it a grip puller. It's all of these and more. Tractel, the world's largest manufacturer, officially calls their hand winches a *griphoist-tirfor-greifzug* product. In English, the word "griphoist" says it all. But the tool was originally sold as a *tirfor*, which in French says much the same. (*Tir* from the French for pull. *For*, probably from *fort*, strong or powerful.) *Greifzug* is the German equivalent. (*Greif* for taking hold or gripping, and *zug* for pulling.)

Lowering a wind turbine for service. *Hand-operated griphoists may revolutionize the installation and servicing of micro and mini wind turbines. Griphoists give much greater control over the raising and lowering of tilt-up towers than either a truck or tractor. Here the Folkecenter for Renewable Energy's Niels Ansø uses a car to anchor a griphoist for lowering a Whisper H1300.*

Lowering a wind turbine for service. *Niels Ansø (background) signals as Folkecenter interns Felix Varela (Cuba) and Hallgrimur Halldorsson (Iceland) guide the tower to its stand.*

A griphoist is "hard to beat for erecting tilt-up towers," says Piggott, "because it is slow and fail-safe." Unlike when a truck or other vehicle is used to raise a tower, the operator of the winch has full control of the operation, and there's no dependence on hand signals or risk of missed cues.

Another hoist option, one used by some American meteorologists to install anemometer masts, is an electric winch. They typically power the winch from the battery of a truck or haul in special-purpose batteries. But Salmon prefers a griphoist to raise NRG tow-

ers in Canada. "They [griphoists] are easier to control," than either winches or vehicles, he says, "and in some cases much safer than [electric] winches."

Unlike electric winches, griphoists are readily portable. You can lug a griphoist into areas where you would never consider hauling an electric winch and battery—or even driving a truck, for that matter.

A griphoist is not a "come-along," a lightweight tool found in North American hardware stores that uses a spool for coiling a short length of wire rope. Ranchers, for example, use come-alongs to tighten fencing. And for that purpose they don't need a long cable.

It's the spool, or drum, that sets come-alongs, as well as winches in general, apart from griphoists. Technically, griphoists are not winches but hoists. Winches use a drum to spool the hoisting cable, like the large drum on a crane. Griphoists, in contrast, pull the hoisting cable directly through the body of the hoist. Tractel likens the locking cams inside the griphoist with the way we take in a rope, "hand over hand." To use a griphoist, you move a lever forward and back. This lever pulls the cable through the tool.

The hoisting cable for a griphoist can be any length, since there is no need to spool the cable on a drum. (Capstan winches can also use cables of any length, but they pass the cable over a drum.)

Like come-alongs, griphoists can float between the load and the anchor for the hoist. Electric winches and hand-cranked mechanical winches are all intended to be mounted on something solid, like a boat deck or the frame of a sport-utility vehicle. (Automotive winches found on some SUVs may have sufficient cable on the drum to raise towers.)

Griphoists and Electric Wire Rope Hoists (Winches).

Type	Model	Capacity		Rope Diameter		Weight		Rope length		Price
		(lbs)	(kg)	(in)	(mm)	(lbs)	(kg)	(ft)	(m)	(1998$)
Griphoist	Pull-All	700	300	3/16	4.75	3.9	1.8	32.8	10	$115
Griphoist	Super Pull-All	1,500	700	1/4	6.30	8.3	3.8	32.8	10	$390
Griphoist	T-508	2,000	900	5/16	8.30	14.5	6.6	32.8	10	$495
Griphoist	T-516	4,000	1,800	7/16	11.50	30.0	13.5	32.8	10	$679
Griphoist	T-532	8,000	3,600	5/8	16.30	51.0	24.0	32.8	10	$1,170
NRG winch	X1*	1,500	907	1/8	3.18	27.0	12.3	165.0	50	$475
NRG winch	S6*	6,000	272	1/4	6.35	128.0	58.0	165.0	50	$925

Does not include cost or weight of battery, but does include pulley blocks and battery cables.

CHOOSING A GRIPHOIST

There are several brands of griphoists on the market, but they all pull a steel cable through the body of the hoist. They come in a range of capacities suitable for most small wind turbine applications. Either the wind turbine or the tower manufacturer will specify the capacity in pounds (or kilograms) needed to raise a tilt-up tower of a given height.

Tractel offers the Pull-All as an entry-level griphoist. It's inexpensive, but it has a serious drawback: Neither the hook on the hoisting cable nor the hook on the body of the griphoist has safety keepers. You can never predict what may happen when you're raising a load, and often there are some jerky movements despite your best efforts. Safety keepers, or latches, keep the hooks engaged when there's unintended slack in the cable, and are absolutely essential.

Tractel's Super Pull-All, the Pull-All's bigger brother, is a real tool. The Super Pull-All weighs twice as much as the Pull-All and it has twice the working load. It also comes with safety keepers on both forged hooks. The Super Pull-All isn't cheap—nearly $400—

but then again, good tools never are. It's shipped with 10 meters (32.8 feet) of 1/4-inch wire rope and two wire rope slings. You can order a longer cable if you need it. Tractel also makes three other sizes.

While little has been written about griphoists, it's surprising the number of people who have used or are now using them. Bergey Windpower, for example, has been using griphoists for remote installations since 1993, when they first used one to raise a 10 kW Excel on an offshore platform. They recommend griphoists to their overseas clients, says Pieter Huebner, Bergey's field technician. One manufacturer, Vergnet, offers griphoists as an accessory to their tower kits. If you have to buy any one tool for your off-the-grid wind system, says Hugh Piggott, it should be a griphoist.

TOWER CONDUCTORS

Never hang the conductors, the power cables that run between the wind turbine and the ground, from the wind turbine leads (small wires). The weight of the cables will

Strain relief. *Drilling a hole for attaching a strain relief (left). Demonstration of how the strain relief is bolted to the inside of an NRG mast for supporting the weight of conductors (power cables) from the wind turbine (right).*

eventually pull the leads out of the generator, causing big headaches later.

On guyed tubular towers, thread the power cables down the inside of the mast. Support the weight of the conductors with a strain relief wire net. The net works like a Chinese finger puzzle to grip the cable bundle. Hang the strain relief from an attachment point at the top of the tower. The strain relief can make or break a wind turbine installation. Many homeowners have yanked the leads out of their wind generators by overlooking this.

Use in-line compression connectors for making vibration-proof connections between the wind turbine leads and the cables carrying power down the tower. These connectors are often used for connecting the power sup- submersible well pumps, and can be

found in farm supply and plumbing stores. Pros such as Mick Sagrillo and Hugh Piggott warn against using split bolts, a common alternative, as they can work loose from the vibrations that are a normal part of a wind turbine's operation.

When you buy the power cable for inside the hinged tubular towers of Westwind Turbines, the Australian company includes end terminations and the suspension strap necessary to support the power cable. That would make even a hard-nosed buyer like Sagrillo happy. Sagrillo argues that manufacturers should ship termination kits and cable straps with their turbines to ensure that the job gets done right the first time. Only Westwind does. But France's Vergnet offers components with its turbines and towers that most other manufacturers leave out. Vergnet includes 35 meters (115 feet) of power cable with all its turbines. They also include junction boxes, grounding, and lightning protection with all their tower kits.

MAINTENANCE

Even some of the old turbines from the 1970s and early 1980s, when well maintained, will continue to spin out the kilowatt-hours. USDA's old Enertech E44 at Bushland, Texas, has been operating since 1982, and has logged more than 100,000 hours of operation.

You should inspect the turbine and tower at least twice each year, says Mick Sagrillo: once in the spring after the turbine has withstood winter storms, and once in the fall in preparation for winter.

RAISING A SMALL TURBINE

To test the difficulties that one might encounter when raising a small wind turbine, Nancy Nies (my wife) and I decided to install a Bergey 850 on a nearly inaccessible site in southern California's Tehachapi Mountains. Since our purpose was experimentation, we wanted a system that would allow us to raise and lower the turbine with as few people as possible. NRG's tilt-up tower and a griphoist seemed like an ideal combination.

Considering the site and our inexperience, we chose NRG's 64-foot (19.5 meter) tower. We thought that NRG's shorter tower wasn't tall enough to clear nearby trees, and that NRG's taller towers were more than we wanted to handle.

NRG's towers were originally developed as meteorological masts. With the introduction of the Bergey 850 in the mid-1990s, NRG

adapted its tower system to small turbines. The BWC 850 was designed specifically for use with NRG's 4.5-inch (114 mm) mast.

To use NRG's tower system, the hoist or hoisting tackle must be anchored directly below the gin pole when the tower is fully upright. If the hoisting anchor is further from the tower base than the length of the gin pole, the sections could come apart and endanger the lift. (NRG provides a safety cable to prevent this from happening, but no one wants to tempt fate.)

The 64-foot (19.5 meter) tower uses a 20-foot (6 meter) long gin pole comprised of two 10-foot (3.3 meter) sections. Thus, the 64-foot tower requires a hoisting anchor 20 feet (6 meter) from the base of the tower. We attached the griphoist to the lifting anchor with one of the wire rope slings provided with the griphoist.

When raising a tower with a gin pole, one of the first challenges is raising the gin pole itself. We attached the hoisting cable to the top of the gin pole with a shackle, first using the griphoist to raise the gin pole upright. Then we slowly raised the tower, literally inch by inch.

While I operated the griphoist, Nancy kept tension in the rear guy cable with a tag line. She stood well clear of the fall zone at a right angle to the tower.

The griphoist pulls a few inches of cable on each stroke of the rear hoist lever, both on the back stroke and on the forward stroke. Because it's a simple mechanical device, you can actually feel the tension in the cable. This gives the operator a tactile sense of the load. The heavier the load, the more difficult it is to move the lever. The loads in tower raising are greatest when the tower is on the ground, and

1. Screw anchor. *There are three elements that make the NRG tower easy to install: screw anchors, mast sections that slip together, and swaged fittings. Where suitable, hand-driven screw anchors greatly simplify the installation of guyed towers for micro and mini wind turbines. Household-size wind turbines will require power-driven screw anchors or conventional concrete anchors.*

least as the tower nears the vertical. Operating the griphoist takes the most effort when the tower first begins to leave the ground.

We spent a whole day raising the tower without the wind turbine, to familiarize ourselves with the process. Unlike traditional towers with anchors at exact positions relative to the tower, NRG's towers were designed for quick installation under field conditions. To allow for misalignment of the anchors, the guy cables are tensioned by hand. The NRG system doesn't use preformed wire grips or turnbuckles.

In our case the anchors were at different elevations. These misalignments cause tension in the cables to vary during the lift. The thin-walled tubing used on the NRG towers easily buckles. So it's necessary to adjust cable tension as the tower is being raised and lowered. If everything is perfect this isn't necessary, but our site was far from perfect.

After we practiced plumbing the tower to ensure that it was vertical, we lowered the tower. The griphoist has two levers, one for pulling in cable, and one for letting it out. It's easy to let the tower down with the griphoist. You simply use the forward lever to operate the hoist in reverse and pay out cable.

The next day we threaded the conductors through the tower and suspended them from the strain relief, terminated the connections between the turbine and the tower conductors (power cables), mounted the turbine, hung the tail vane, and bolted on the blades.

We then repeated the raising sequence. The 87-pound (39 kg) turbine added 60 percent to the lift load and it could be clearly felt in the griphoist. It took a lot more effort than raising the tower without the turbine, but we raised the turbine and tower in less than one hour. It took another hour to plumb the tower and tighten the guy cables in a stiff wind. Though it wasn't a stroll in the park, physically operating the griphoist during the early part of the lift wasn't difficult. It became much easier once the tower reached about 45 degrees.

After the tower was upright and the Bergey began whirring, Nancy said, "I thought there was going to be a lot more to it than that. It was simpler than I thought." That was the whole idea.

2. Driving the screw anchor. *The anchors with the NRG tower can be driven by hand. Driving five of these anchors into medium-dense soil is enough exercise for one day.*

3. Slip Fit. *NRG tower sections slip together, eliminating bolted connections. The sections seat firmly together when the tower is first raised.*

4. Unreeling guy cables. *The NRG tower is packaged with four reels of guy cable per guy level. All attachments to the guy bracket are swaged. This greatly simplifies assembly.*

5. Gin-pole bracket. *Hoisting cable (left) is independent of lifting guys (right). Nylon rope steadies gin pole during lift. Note lifting guys are clearly labeled.*

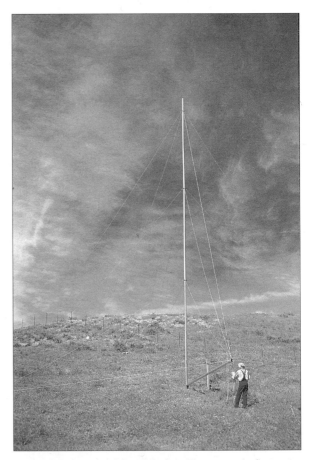

6. Raising tower with griphoist. *The tower is first raised without the wind turbine.*

7. Griphoist. *Using Tractel's Super Pull-All to lower the tower after initial tower raising. Hoisting cable passes through the body of the griphoist. Note safety keeper or gate on griphoist hook.*

8. Lowering tower with griphoist. *Forward lever is used to pay out cable and lower the tower.*

9. Adjusting cable tension. *The NRG system doesn't use turnbuckles for tensioning cables. This allows cable length to be quickly adjusted while raising and lowering the tower.*

10. Tightening wire rope clips. *NRG provides both the wire rope clips and cables to the guy anchors. Remember to "never saddle a dead horse," that is, the forged saddle is placed over the live cable.*

11. Raising the assembled turbine and tower. *The pace is slow, allowing for ample time to check and recheck cables and anchors.*

12. Not quite there. *With the griphoist there is plenty of time to check the plumb of the tower and take in or pay out cable as needed.*

13. Plumbing tower. *Checking plumb with a torpedo level. The turbine will yaw freely when the tower is plumb (vertical).*

14. New American Gothic. *By using the griphoist, two people were able to raise a BWC 850 safely on a difficult hillside site.*

9
Conclusion

In this book we've examined how wind energy is being used to meet various applications, from charging electric fences to feeding power into the utility grid. We've explored the basics of wind energy, including the importance of wind speed and rotor diameter in determining how much energy a wind turbine will capture. We've learned about the differences in technology among small wind turbines and how these machines are used in both off-the-grid power systems and in utility-interconnected applications. And finally, we've discussed where it's best to install them, how to install them safely, what kinds of towers are best suited for the innovative micro and mini wind turbines now on the market, and what it all will likely cost in the end.

The new millennium holds vast promise. The worldwide demand for small wind turbines has never been greater. As the small wind turbine industry grows to meet this need, we can expect the introduction of new products, as well as gradual improvements in reliability and cost-effectiveness.

NREL's Trudy Forsyth says 24 percent of the U.S. population lives in rural areas. This simple observation suggests enormous potential for the use of both hybrid off-the-grid and grid-intertie systems in the United States. In Alaska alone, more than two hundred villages depend upon diesel engines for power.

On the Great Plains of the lower forty-eight states, some utilities are discovering that it's cheaper to install and maintain hybrid power systems than to maintain miles of lightly-used distribution lines, says USDA's Nolan Clark. He estimates it costs $300 per kilometer ($500 per mile) to maintain a rural distribution line. For that price, the utility can service an entire hybrid system, including the wind turbine. Over time, Clark expects this will spur utility interest in end-of-the-line applications.

Such distributed generation has already taken unusual forms. In 1995 for example, Magic Circle Energy Co. installed four Bergey Excels in Oklahoma. Magic Circle, an independent producer, is using the wind turbines to pump oil from its stripper wells. In the foothills of the Pyrenees, a French farmer uses a Vergnet turbine to refrigerate his cheese.

Growing populations and growing aspirations in the Third World are signaling a steadily swelling demand for small wind turbines. These machines will become part of hybrid power systems for village electrification, obviating the need for strapped central governments to extend power lines to every village.

> These and several other encouraging trends offer the tantalizing prospect that small wind turbines could eventually fulfill the vision wind energy advocates once had for them.

The demand for turbines to supply these applications will continue to put pressure on the manufacturers of small wind turbines to improve reliability, ease of installation, and service.

In North America, the increasing number of states and provinces with net-metering regulations is preparing the way for a time when the political winds may well change direction, making intertied small wind turbines economically attractive once again.

Net metering, says attorney Tom Starrs, is one of those issues that resonates with the public and the politicians who presumably serve them. Starrs—net metering's most artic-

ulate proponent—has watched the issue find growing acceptance. And though small interconnected wind turbines are financially unappealing under today's conditions in North America, this situation could change overnight.

Technology is inherently political, say high-tech pundits in the Silicon Valley, and the popularity of net metering, coupled with political interest in at least appearing to tackle global environmental problems, could lead to surprising political action. Germany's highly successful Electricity Feed Law resulted from just such a combination of political opportunism and pragmatism.

These and several other encouraging trends offer the tantalizing prospect that small wind turbines could eventually fulfill the vision wind energy advocates once had for them.

The trend toward certification and standardized testing of small wind turbines will eliminate some of the risk consumers now bear when buying these products. Formal product testing, a facet of wind turbine marketing long ago established for commercial wind turbines, will also pressure manufacturers to quickly upgrade their products to the quality expected of today's consumer commodities. This includes the trend on the part of small wind turbine manufacturers to upgrade peripheral components, such as control panels, to meet the requirements of electrical standards associations. Compliance with these standards, and certification, will greatly simplify acceptance by utilities of interconnected wind turbines and by building inspectors of off-the-grid power systems.

Another trend that bodes well for wind en-

Wind power plant. *Ecotecnia's Galicia wind plant in northwestern Spain, one of the country's most attractive. Ecotecnia's 26-meter diameter, 250-kilowatt turbines produce commercial quantities of wind-generated electricity.*

ergy is the increasing concern about global climate change. Programs intended to address this issue by intensifying development of renewable energy will also likely benefit small wind turbines.

The tremendous growth in commercial wind power will undoubtedly spur advancements in small wind turbines, albeit indirectly. The prominence of commercial wind turbines in many parts of the world yields an increased public awareness of wind energy in general. Developments in commercial wind turbines also result in a better understanding of wind technology and how to make it work more reliably. The technological spin-offs from medium-size wind turbines, combined with the financial acumen that gravitates toward a billion-dollar industry, could breathe new life into an old, but still exciting technology.

Appendix 1

Characteristics of Selected Small Turbines

Selected Micro Turbines.

Manufacturer	Model	Rotor Diameter (m)	(ft)	Area (m²)	Mfg. Rated Power (kW)	Rated Wind Speed (m/s)
Marlec	Rutland 500	0.51	1.7	0.20	0.02	10.0
LVM	Aero2gen	0.58	1.9	0.26	0.01	10.3
LVM	Aero4gen F	0.86	2.8	0.58	0.06	10.3
Marlec	Rutland 913	·0.91	3.0	0.65	0.07	10.0
Ampair	100	0.92	3.0	0.66	0.05	10.0
Southwest Windpower	Air 403	1.20	3.8	1.07	0.40	12.5
Aerocraft	AC 120	1.20	3.9	1.13	0.12	9.0
LVM	Aero6gen F	1.20	4.0	1.17	0.12	10.3

Selected Mini Turbines.

Manufacturer	Model	Rotor Diameter (m)	(ft)	Area (m²)	Mfg. Rated Power (kW)	Rated Wind Speed (m/s)
Southwest Windpower	Windseeker 503	1.5	5.0	1.81	0.50	12.5
World Power Tech.	H500	1.5	5.0	1.83	0.50	12.5
LVM	Aero8gen F	1.6	5.1	1.89	0.22	10.3
Aerocraft	AC 240	1.7	5.4	2.14	0.24	9.0
Marlec	Rutland 1800	1.8	5.9	2.54	0.25	9.4
World Power Tech.	H900	2.1	7.0	3.58	0.90	12.5
Aerocraft	AC 500	2.4	7.9	4.52	0.50	8.5
Aerocraft	AC 750	2.4	7.9	4.52	0.75	9.5
Proven Wind Turbines	WT600	2.6	8.4	5.11	0.60	10.0

Legend

nc = no control
h = horizontal furling
v = vertical furling
ab = air brake
p = pitch to feather

ps = pitch to stall
pt = pitchable tip
mb = mechanical brake
GFRP = glass fiber reinforced polyester or fiberglass
wood lam. = wood laminate

CFRP = carbon fiber reinforced polyester
CFRE = carbon fiber reinforced epoxy
Poly = glass fiber reinforced polypropylene
PGFR = polycarbonate glass fiber reinforced

No. of Blades	Blade Material	Means of Control	Mass (kg)	Spec. Mass (kg/m²)	Model	Manufacturer
6	nylon	nc			Rutland 500	Marlec
5	nylon	nc	5.0	19	Aero2gen	LVM
6	nylon	h	11.0	19	Aero4gen F	LVM
6	nylon	nc	10.5	16	Rutland 913	Marlec
6	Poly	nc	12.0	18	100	Ampair
3	CFRP	nc	5.9	6	Air 403	Southwest Windpower
5	GFRP	v	17.0	15	AC 120	Aerocraft
6	nylon	h	16.0	14	Aero6gen F	LVM

No. of Blades	Blade Material	Means of Control	Mass (kg)	Spec. Mass (kg/m²)	Model	Manufacturer
3	wood	v	10.2	6	Windseeker 503	Southwest Windpower
3	PGFR	v	13.0	7	H500	World Power Tech
3	wood lam.	h	18.0	10	Aero8gen F	LVM
5	GFRP	v	19.0	9	AC 240	Aerocraft
3	GFRP	h			Rutland 1800	Marlec
3	PGFR	h	25.0	7	H900	World Power Tech
3	GFRP	p	41.0	9	AC 500	Aerocraft
3	GFRP	p	43.0	10	AC 750	Aerocraft
3	Poly	ps	65.0	13	WT600	Proven Wind Turbines

Selected Household-size Turbines.

Manufacturer	Model	Rotor Diameter (m)	(ft)	Area (m²)	Mfg. Rated Power (kW)	Rated Wind Speed (m/s)
World Power Tech.	1000	2.7	9.0	5.91	1.0	11.0
Bergey Windpower	1500-24	3.1	10.0	7.31	1.5	12.5
Proven Wind Turbines	WT2500	3.5	11.5	9.62	2.2	12.0
Westwind	Light Wind	3.6	11.8	10.20	1.8	12.5
Westwind	Standard	3.6	11.8	10.20	2.5	14.0
Vergnet	GEV 4.1	4.0	13.1	12.60	1.1	8.0
World Power Tech.	3000	4.5	14.8	16.00	3.0	11.0
Calorius		5.0	16.4	19.60		
Vergnet	GEV 5.5	5.0	16.4	19.60	5.0	13.0
Aerocraft	AC 3000	5.3	17.2	21.60	3.0	10.0
Proven Wind Turbines	WT6000	5.5	18.0	23.80	6.0	10.0
Vergnet	GEV 6.5	6.0	20.0	28.30	5.0	10.0
Vergnet	GEV 7.6	7.0	23.0	38.50	6.0	10.0
Vergnet	GEV 7.10	7.0	23.0	38.50	10.0	11.5
Bergey Windpower	Excel PD	7.0	23.0	38.50	10.0	12.1
Bergey Windpower	Excel R/48	7.0	23.0	38.50	6.5	12.1
Bergey Windpower	Excel-S	7.0	23.0	38.50	10.0	12.1
Westwind	Light Wind	7.0	23.0	38.50	8.0	12.5
Westwind	Standard	7.0	23.0	38.50	10.0	13.5
Wind Turbine Ind.	23-10	7.0	23.0	38.60	10.0	11.6
Wind Turbine Ind.	23-12.5	7.0	23.0	38.60	12.5	12.5
Wind Turbine Ind.	29-20	8.8	29.0	61.40	20.0	11.6

Legend

nc = no control
h = horizontal furling
v = vertical furling
ab = air brake
p = pitch to feather

ps = pitch to stall
pt = pitchable tip
mb = mechanical brake
GFRP = glass fiber reinforced polyester or fiberglass
wood lam. = wood laminate

CFRP = carbon fiber reinforced polyester
CFRE = carbon fiber reinforced epoxy
Poly = glass fiber reinforced polypropylene
PGFR = polycarbonate glass fiber reinforced

No. of Blades	Blade Material	Means of Control	Mass (kg)	Spec. Mass (kg/m²)	Model	Manufacturer
2	CFRE	h	30	5	1000	World Power Tech
3	GFRP	h	76	10	1500-24	Bergey Windpower
3	Poly	ps	190	20	WT2500	Proven Wind Turbines
3	GFRP	h			Light Wind	Westwind
3	GFRP	h			Standard	Westwind
2	wood lam.	ps			GEV 4.1	Vergnet
2	CFRE	h	70	4	3000	World Power Tech.
3	GFRP					Calorius
2	wood lam.	ps			GEV 5.5	Vergnet
3	GFRP	p	190	9	AC 3000	Aerocraft
3	Poly	ps	400	17	WT6000	Proven Wind Turbines
2	wood lam.	ps			GEV 6.5	Vergnet
2	wood lam.	ps			GEV 7.6	Vergnet
2	wood lam.	ps			GEV 7.10	Vergnet
3	GFRP	h	463	12	Excel PD	Bergey Windpower
3	GFRP	h	463	12	Excel R/48	Bergey Windpower
3	GFRP	h	463	12	Excel-S	Bergey Windpower
3	GFRP	h			Light Wind	Westwind
3	GFRP	h			Standard	Westwind
3	wood lam.	p,h	636	16	23-10	Wind Turbine Ind.
3	wood lam.	p,h			23-12.5	Wind Turbine Ind.
3	GFRP	p,h	1045	17	29-20	Wind Turbine Ind.

Appendix 2

Resources

SELECTED SMALL WIND TURBINE MANUFACTURERS

There are more than fifty manufacturers of small wind turbines worldwide, and they produce more than one hundred different models. Only the selected few presented in the text are listed here. For additional addresses contact the wind energy associations listed in the following section.

Aerocraft
Hoffeldstrasse 20
D27356 Rotenburg
Germany
phone: 49 42 61 96 00 34
fax: 49 42 61 96 00 35

AmpAir
P.O. Box 416
Poole Dorset BH12 3LZ
United Kingdom
phone: 44 12 02 74 99 94
fax: 44 12 02 73 66 53
email: ampair@ampair.com
www.ampair.com

Bergey Windpower Co.
2001 Priestley Ave
Norman, OK 73069
phone: 405 364 4212
fax: 405 364 2078
email: sales@bergey.com
www.bergey.com

Calorius Vindmøllen
Spanagervej 2
DK 4200 Slagelse Sjaelland
Denmark
phone: 45 58 26 80 60
fax: 45 58 26 80 60
email: calorius@post7.tele.dk

LVM Ltd.
Old Oak Close
Arlesey Bedfordshire SG15 6XD
United Kingdom
phone: 44 14 62 73 33 36
fax: 44 14 62 73 04 66
email: lvmltd@dial.pipex.com

Marlec Engineering Co.
Rutland House
Trevithick Road
Corby Northants NN17 1XY
United Kingdom
phone: 44 15 36 20 15 88
fax: 44 15 36 40 02 11
email: sales@marlec.co.uk
www.marlec.com

Proven Engineering Products
Moorfield Industrial Estates
Kilmarnock Scotland KA2 0BA
United Kingdom
phone: 44 15 63 543 020
fax: 44 15 63 539 119
email: gordon.proven@post.almac.co.uk
www.almac.co.uk/proven/

Southwest Windpower
2131 N. First St.
P.O. Box 2190
Flagstaff, AZ 86003-2190
phone: 520 779 9463
fax: 520 779 1485
email: info@windenergy.com
www.windenergy.com

Vergnet S.A.
6 rue Henri Dunant
F 45140 Ingre
France
phone: 33 2 38 22 75 00
fax: 33 2 38 22 75 22
email: vergnet@wanadoo.fr
www.vergnet.fr/

Westwind Turbines
29 Owen Rd.
Kelmscott, Western Australia 6111
Australia
phone: 61 8 93 99 52 65
fax: 61 8 94 97 13 35
email: venwest@iinet.net.au
www.venwest.iinet.net.au/

Wind Turbine Industries
16801 Industrial Circle, SE
Prior Lake, MN 55372
phone: 612 447 6064
fax: 612 447 6050

World Power Technologies
19 Lake Ave. North
Duluth, MN 55802
phone: 218 722 1492
fax: 218 722 0791
email: sales@worldpowertech.com
www.worldpowertech.com

GUYED TOWERS FOR MICRO AND MINI WIND TURBINES

NRG Systems Inc.
110 Commerce St.
P.O. Box 509
Hinesburg, VT 05461
phone: 800 448 9463, or 802 482 2255
fax: 802 482 2272
email: sales@nrgsystems.com
www.nrgsystems.com

RECONDITIONED SMALL TURBINES

Used and reconditioned small wind turbines can be found in the classifieds of *Home Power* magazine or by contacting Lake Michigan Wind & Sun.

Lake Michigan Wind & Sun
1015 County U
Sturgeon Bay, WI 54235-8353
Phone: 920 743 0456
Fax: 920 743 0466
email: lmwands@itol.com

SELECTED MANUFACTURERS AND DISTRIBUTORS OF WIND PUMPS

Aermotor Windmill Corp.
Box 5110
San Angelo, TX 76902
phone: 800 854 1656 or 915 651 4951
fax: 915 651 4948
email: info@aermotorwindmill.com
www.aermotorwindmill.com

Bowjon International
11580 Lawton Court
Loma Linda, CA 92354
phone: 909 796 7199
email: bowjon@gte.net
http://home1.gte.net/bowjonin/indx.htm

Dempster Industries
P.O. Box 848
Beatrice, NE 68310
phone: 402 223 4026
fax: 402 228 4389
email: dempsterinc.@beatricene.com

Heller-Aller
P.O. Box 29
Napoleon, OH 4345
phone: 419 592 1856

Koenders Windmills
P.O. Box 126
Englefeld, Saskatchewan S0K 1N0
Canada
phone: 306 287 3702
fax: 306 287 3657
email: koenders.wind@sk.sympatico.ca
www.humboldtsk.com/Koenders/index.html

O'Brock Windmill Distributors
9435 12th St.
North Benton, OH 44449
phone: 330 584 4681
fax: 330 584 4682
email: windmill@cannet.com
(send $3.00 for catalog)

Southern Cross Corp.
632 Ruthven St.
P.O. Box 109
Toowoomba, Queensland QLD 4350
Australia
phone: 61 76 38 4988
fax: 61 76 38 5898

WIND MEASUREMENT DEVICES

Ammonit
Paul-Lincke-Ufer 41
D10999 Berlin
Germany
phone: 49 30 612 79 54
fax: 49 30 618 30 60
email: ammonit@ammonit.de
www.ammonit.de

NRG Systems Inc.
110 Commerce St.
P.O. Box 509
Hinesburg, VT 05461
phone: 800 448 9463, or 802 482 2255
fax: 802 482 2272
email: sales@nrgsystems.com
www.nrgsystems.com

R.M. Young Co.
2801 Aero Park Dr.
Traverse City, MI 49684
phone: 616 946 3980
fax: 916 946 4772
email: met.sales@youngusa.com
www.youngusa.com

Second Wind
366 Summer St.
Somerville, MA 02144
phone: 671 776 8520
fax: 617 776 0391
email: sales@secondwind.com
www.secondwind.com

INVERTERS AND CONTROLS

Advanced Energy Systems
14 Brodie Hall Drive
Bentley, Western Australia 6102
Australia
phone: 64 8 94 70 46 33
fax: 64 8 94 70 45 04
email: info@advancedenergy.com
www.advancedenergy.com/company/

Pulse Energy Systems
870-E Gold Flat Road
Nevada City, CA 95959
phone: 916 265 9771
fax: 916 265 9756
email: info@pulseenergy.com
www.pulseenergy.com

Selectronic Australia
25 Holloway Dr.
Bayswater Victoria 3153
Australia
phone: 61 3 9762 4822
fax: 61 3 9762 9646
email: sales@selectronic.com.au
www.selectronic.com.au

Trace Engineering
5916 195th NE
Arlington, WA 98223
phone: 360 435 8826
fax: 360 435 2229
email: inverters@traceengineering.com
www.traceengineering.com

Vanner Power Group
4282 Reynolds Dr.
Hilliard, OH 43026
phone: 614 771 2718
fax: 614 771 4904
email: pwrsales@vanner.com
www.vanner.com

GRIPHOISTS

Tractel (USA)
P.O. Box 707
Westwood, MA 02090
phone: 781 329 5650
fax: 781 329 6350
email: griphoist@worldnet.att.net

Tractel (World)
85–87, rue Jean Lolive
F93100 Montreuil
France
phone: 33 1 48 58 91 32
fax: 33 1 48 58 19 95

WIND ENERGY ASSOCIATIONS

For more information about how to use wind energy in your country, contact the following wind energy organizations:

American Wind Energy Association (AWEA)
122 C St. NW, 4th Fl
Washington, DC 20001
phone: 202 383 2500
fax: 202 383 2505
email: windmail@awea.org
www.awea.org

Association of Danish Windmill Manufacturers
(Dansk Vindmølleindustrien)
Vester Voldgade 106
DK 1552 Copenhagen
Denmark
phone: 45 33 73 03 30
fax: 45 33 73 03 33
email: danish@windpower.dk
www.windpower.dk

British Wind Energy Association (BWEA)
26 Spring St.
London W2 1JA
United Kingdom
phone: 44 171 402 7102
fax: 44 171 402 7107
email: bwea@gn.apc.org
www.bwea.com

Canadian Wind Energy Association (CanWEA)
100, 3553 31st NW
Calgary, Alberta T2L 2K7
Canada
phone: 800 9 CanWEA
fax: 403 282 1238
email: canwea@canwea.com
www.canwea.ca

Dutch National Bureau for Wind Energy (Landelijk Bureau Windenergie)
Postbus 10
NL 6800 AA Arnhem
The Netherlands
phone: 31 26 37 79 700
fax: 31 26 37 79 707
email: lbwinfo@windenergy.nl
www.windenergy.nl

European Wind Energy Association (EWEA)
26 Spring St.
London W2 1JA
United Kingdom
phone: 44 171 402 7122
fax: 44 171 402 7125
email: ewea@compuserve.com

German Wind Energy Association
(Bundesverband WindEnergie, BWE)
Natruper Strasse 70
Osnabrück D 49090
Germany
phone: 49 541 96 19 185
fax: 49 541 96 19 186
email: BWE_Os@t-online.de
www.wind-energie.de

SOURCES OF WIND DATA

Deutsches Windenergie Institut (DEWI)
Eberstrasse 96
D2940 Wilhelmshaven
Germany
phone: 49 4421 48080
fax: 49 4421 480843
email: dewi@dewi.de
www.dewi.de

Energy Technology Support Unit (ETSU)
B156 Harwell Laboratory
Didcot, Oxfordshire OX11 ORA
United Kingdom
phone: 44 12 35 43 35 17
fax: 44 12 35 43 29 23

Institut für Solare Energieversorgungstechnik (ISET)
Königstor 59
D34119 Kassel
Gemany
phone: 49 561 72 940
fax: 49 561 72 94 100
www.iset.uni-kassel.de/

National Renewable Energy Laboratory/NWTC
1617 Cole Blvd.
Golden, CO 80401-3393
phone: 303 384 6900
fax: 303 384 6999
www.nrel.gov/wind/

MAIL-ORDER CATALOGS

Many micro and mini wind turbines can be purchased from mail-order companies. The turbines are small enough to be shipped via package delivery services, such as United Parcel Service (UPS). The following companies also sell the heavier small wind turbines, but they ship them by motor freight (truck). For wind turbines 1.5 kW and above, it is often best to buy them from a local dealer who can both install and service the machine.

Alternative Energy Engineering
P.O. Box 339
Redway, CA 95560-0339
phone: 800 777 6609, or 707 923 2277
fax: 707 923 3009
email: energy@alt-energy.com
www.alt-energy.com

Jade Mountain
P.O. Box 4616
Boulder, CO 80306
phone: 800 442 1972, or 303 449 6601
fax: 303 449 8266
email: info@jademountain.com
www.jademountain.com

Kansas Wind Power
13569 214th Road
Holton, KS 66436
phone: 913 364 4407
fax: 913 364 4407

Real Goods Trading Co.
555 Leslie St.
Ukiah, CA 95482-5507
phone: 800 919 2400, or 707 468 9214
fax: 707 462 4807
email: techs@realgoods.com
www.realgoods.com

PUBLICATIONS

Plans

Centre for Alternative Technology
Llwyngwern Quarry
Machynlleth Powys, Wales SY20 9AZ
United Kingdom
phone: 44 16 54 702 400
fax: 44 16 54 702 782
email: help@catinfo.demon.co.uk
www.cat.org.uk

Jade Mountain
P.O. Box 4616
Boulder, CO 80306
phone: 800 442 1972
fax: 303 449 8266
email: info@jademountain.com
www.jademountain.com

Kragten Design
Populierenlaan 51
NL 5492 SG Sint-Odenrode
The Netherlands
phone: 31 41 38 75 770
fax: 31 41 38 75 770

Lindsay Publications
P.O. Box 12
Bradley, IL 60915-0012
phone: 815 935 5353
fax: 815 935 5477
email: lindsay@lindsaybks.com
www.lindsaybks.com

Scoraig Wind Electric
Dundonnell, Ross shire
Scotland IV23 2RE
United Kingdom
phone: 44 18 54 633 286
fax: 44 18 54 633 286
email: hugh.piggott@enterprise.net
www.windmission.dk/workshop/

Books

Independent Energy Guide: Electrical Power for Home, Boat & RV, by Kevin Jeffrey. Orwell Cove Press. 1995. Distributed by Chelsea Green Publishing Co., White River Junction,Vermont. ISBN 0-9644112-0-2, 288 pages, $19.95. A comprehensive guide for planning the ideal independent power system for your home, boat, or RV.

Renewables Are Ready: People Creating Renewable Energy Solutions, by Nancy Cole and P. J. Skerrett. Chelsea Green Publishing Co., White River Junction, Vermont. 1995. ISBN: 0-930031-73-3, 254 pages, $19.95. Documents the renewable energy technologies being put to use in diverse communities across the country.

The Solar Electric House: Energy for the Environmentally-Responsive, Energy-Independent Home, by Steven Strong and William Scheller. Sustainability Press. 1994. Distributed by Chelsea Green Publishing Co., White River Junction, Vermont. ISBN 0-963738-32-1, 288 pages, $21.95. Explains in detail how to use photovoltaics to power your home.

Who Owns the Sun: People, Politics, and the Struggle for a Solar Economy, by Daniel M. Berman and John T. O'connor. Chelsea Green Publishing Co., White River Junction, Vermont. 1996. ISBN 1-890132-08-X, 342 pages, $24.95. Shows how existing solar technologies combined with local managment present logical remedies for our energy gluttony.

Windpower Workshop, by Hugh Piggott. Centre for Alternative Technology Publications, Machynlleth, Wales, United Kingdom. May 1997. ISBN 1-898049-13-0, 158 pages, UK £7.95. An ideal introduction to building your own home wind system by someone who knows how to do it.

Wind Power for Home & Business, by Paul Gipe. Chelsea Green Publishing Co., White River Junction, Vermont. 1993. ISBN 0-930031-64-4, 414 pages, $35. Describes how homeowners, farmers, and small businesses can use small and medium-size wind turbines. Detailed and far more comprehensive than this text.

Wind Energy Comes of Age, by Paul Gipe. John Wiley & Sons, New York. 1995. ISBN 0-471-10924-X, 536 pages, $70. An overview of the commercial development of wind energy worldwide.

Magazines

Home Power is the only magazine that regularly covers the practical aspects of using small wind turbines at remote or off-the-grid homesteads. In Australia both *ReNew* and *Earth Garden* cover remote power systems.

Earth Garden
RMB 427
Trentham, Victoria 3458
Australia
phone: 61 3 5424 1819
fax: 61 3 5424 1743
email: alangray@kyneton.net.au
www.earthlink.com.au/earthgarden

Home Power Magazine
P.O. Box 520
Ashland, CA 97520
phone: 530 475 3179
fax: 530 475 0836
email: hp@homepower.com
www.homepower.com

ReNew
P.O. Box 2001
Lygon Street North
Brunswick, East Victoria
Australia
phone: 61 3 9650 7883
fax: 61 3 9650 8574

Market Surveys

The German Wind Energy Association annually publishes a comprehensive survey of wind turbine manufacturers worldwide. The survey includes small wind turbines. The text is in both German and English.

Windkraftanlagen Marktübersicht
(Windturbine Market Survey)
Herrenteichsstrasse 1
D49074 Osnabrück
Germany
phone: 49 541 350 6031
fax: 49 541 350 6030
email: ne-bwe@t-online.de
www.wind-energie.de

Videos

Mick Sagrillo and Scott Andrews have produced a helpful video on installing small wind turbines. The video (in NTSC) is also available from AWEA.

An Introduction to Residential Wind Power with Mick Sagrillo, by Scott Andrews, 63 minutes, $39.95 plus $5 postage and handling, P.O. Box 3027, Sausalito, CA 94965; ph: 415 332 5191.

ELECTRONIC INFORMATION SOURCES

The American Wind Energy Association sponsors a news group on small wind turbines. The discussion is often enlightening and can sometimes become quite lively. To subscribe, write Tom Gray (tomgray@awea.org) and ask to be placed on the awea.wind.home list. Both AWEA and the Association of Danish Windmill Manufacturers sponsor informative Web sites. You can find them at www.awea.org

www.windpower.dk

Index

A

Aerocraft, 10, 30, 34
air density, 7–8
air-lift pumps, 48, 50
Alaska, 103
Alberta, Canada, 52
Alternative Energy Institute, 79
alternators, 28–29
American Wind Energy Association, 12, 67
Ampair 100 (turbine), 2, 32
anchors, screw, 96, 97
anemometer masts, 8, 92
anemometers, 9, 112–13
Ansø, Niels, 90, 91
asynchronous generators. *See* induction generators

B

batteries, 44–45, 77
battery charging, 17, 34, 41
Bayly, Elliott, 35, 43, 44, 45
Bergey, Mike, 63
Bergey Windpower
850 (turbine), 35, 36
1500 (turbine), 37, 38, 51, 60
Excel (turbine), 5, 12–14, 38, 56
furling technology, 27, 28
generators, 28
griphoist recommendations, 93
inverters, 64
mast recommendations, 40
power, 20, 21, 23
reliability of, 19
blade
composition, 25–26
length, 9
number, 25
pitching, 27
brake switch, 77
Bushland, Texas, experiment station, 19, 48

C

cabins, 34, 45, 54–57
California, 1, 51, 61, 87
Calley, Dave, 33
Calorius Vindmøllen, 53
Canada, 52, 71–72
capstan winches, 92
Caribbean, 11, 58
Centre for Alternative Technology, 85, 86
China, 2
Clark, Nolan, 25, 48, 83, 84, 87, 103
climbing, 78–79
come-alongs, 92
conductors, 93–95, 96
control panels, 77, 83–84
cottages. *See* cabins

D

Darrieus turbines, 39–40
data loggers, 8, 9
DC Source Center, 45
Denmark, 4, 72
excess power generation, 43
fence charging, 57
interconnected systems, 67–69

Deutches Windenergie Institut, 12
Dorsett, Bill, 43
Duluth, Minnesota, 47
dump loads, 17, 52, 66

E
Edworthy, Jason, 11, 12, 14, 52, 71–72, 81
Eggleston, Eric, 48
electric fences, 29, 31, 55, 57
Electricity Feed Law, 69, 104
electric utility systems. *See* utility interties
electric vehicles, 57
electric winches, 92
ENEL, 59
energy, 10
    excess generation, 17, 52, 66
    output estimation, 17–23, 29–30
    See also wind power
Energy Technology Support Unit, 12
Enertech, 62, 63, 95
Europe, 3–5, 43, 67–69. *See also specific countries*

F
fall-arresting cable, 78–79, 95
farm windmills, 48, 49
fences. *See* electric fences
Folkecenter for Renewable Energy, 45, 50–52
Forsyth, Trudy, 59, 103
France, 59, 104
furling, 26–28, 29

G
generators, 3, 28–29, 63–64
Germany

Electricity Feed Law, 69, 104
    excess power generation, 43, 66–67
gin poles, 89, 98
Great Britain, 38, 52, 73
Green, Jim, 81
griphoists, 98, 99
    advantages, 90–92
    manufacturers, 90, 113
    operation, 92, 95–96
    selection, 93
guy cables, 79, 97
guyed masts, 40, 41
    anemometer masts, 8
    conductors in, 94
    installation, 89–90, 95–101
    lattice, 72, 79
    manufacturers, 83, 111
    placement, 78
    tilt-up, 41, 79, 89

H
heating, 34
    with electricity, 50–52
    with excess generation, 52
    with wind furnace, 50
height
    tower, 14–15, 71–75
    wind speed and, 12–15
horizontal furling, 27, 28
household-size turbines, 2, 4, 5, 13, 36–38, 68
    characteristics, 108–109
    generators, 28
    manufacturers, 38, 110–11
Huebner, Pieter, 93
hybrid wind and solar systems, 44–45, 46
    components, 44–45
    uses, 43, 54, 59

I
incentives, 87
Indonesia, 59
induction generators, 3, 28, 63–64
in-line compression connectors, 94
installation
    griphoist in, 90–93
    on rooftop, 40–41, 54, 73
    safety in, 93, 95
    on tower, 79, 89–90
    on tree, 73–75
    truck/tractor in, 89–90
Institut für Solare Energieversorgungstechnik, 12
integrated-gate, bipolar transistors (IGBTs), 64–65
interconnected systems. *See* utility interties
inverters, 113
    hybrid systems, 44, 45
    utility interties systems, 64–65

J
Jones, Marty, 33

K
Kragten Design, 85
Kruse, Andy, 73
Kuriants, 69

L
La Désirade, 58
Lake Michigan Wind & Sun, 72, 85
lattice towers, 40, 41, 79
LeBeau, Mike, 47
line-communicated inverters, 64
LVM Ltd., 2, 27, 31

M

Magic Circle Energy Co., 103
maintenance, 78–79, 95
manufacturer's estimates,
    21–23
Marlec Engineering Co., 2, 13,
    30, 55
    furling technology, 27
    Rutland 913 (turbine), 30,
        31–32
masts. *See* towers
measurement, units of, 12
Mexico, 59
micro turbines, 2, 12, 13, 16,
    29–34
    characteristics, 106–107
    generators, 28
    manufacturers, 31–34,
        110–11
    uses, 29, 54, 57
mini turbines, 13, 34–36
    characteristics, 106–107
    generators, 28
    manufacturers, 35–36,
        110–11
Minnesota, 47

N

National Renewable Energy
    Laboratory, 12, 34, 59, 81
net metering, 63, 67, 104
New Zealand, 57
Nies, Nancy, 95, 96
noise level, 75
North America, 59, 104. *See
    also* United States
Northwest Territories,
    Canada, 71–72
NRG Systems towers, 59, 89,
    90, 97
    installation, 83, 95–96, 100
    selection, 72–73

O

off-the-grid power systems,
    43–59
    hybrid solar/wind systems,
        43, 44–45, 46
    safety, 77
    uses, 48–59
overspeed control, 26–28

P

Park, Jack, 11
permanent-magnet alterna-
    tors, 28, 29
photovoltaic modules, 12, 43,
    46
Piggott, Hugh, 29, 31, 32, 34,
    38, 52, 73, 76, 78, 90, 92,
    93
power. *See* energy; wind
    power
power cables, 93–95, 96
power curve estimation
    method, 20–21
Pratt, Doug, 25, 29, 45
Proven Engineering, 38
    blade pitching, 27
    down-wind orientation, 26,
        38
    towers, 73
Public Utility Regulatory Poli-
    cies Act (PURPA), 61
Puerto Rico, 11
Pulse Energy Systems, 77
pumps, 31, 48, 112

Q

Quintana Roo, Mexico, 59

R

Rayleigh wind speed distribu-
    tion, 10, 11

recreational vehicles, 29, 45, 54
Renewable Energy Feed-in
    Tariffs (REFITS), 69
repairs, 78–79
Rohn 25G (guyed tower), 72
rooftop mounting, 40–41, 54,
    73
rotor
    blade length, 9
    diameter, 9, 13, 18

S

safety belt and lanyard, 78
safety cable, 78–79, 95
safety keepers, 93
safety procedures
    for batteries, 77
    for electrical systems, 77
    for maintenance, 78–79
    for rotating machinery,
        76–77
    in tower raising, 79, 93, 95
    for towers, 77–78
Sagrillo, Mick, 9, 26, 73, 78,
    89, 90, 94
sailboats, 29, 31, 32, 54
Salmon, Jim, 90, 92
Scoraig Wind Electric, 85
Scotland, 38, 73
screw anchors, 96, 97
self-commutated inverters,
    64–65
silicon controlled rectifier
    (SCR) switches, 64
sine wave inverters, 45, 65
small turbines, 1
    collective purchases, 86–87
    energy output, 19
    generators, 28–29
    installation, 40–41, 73,
        89–101
    kits, 85

maintenance, 78–79, 95
orientation, 25, 26
overspeed control, 26–28
prices, 84
robustness, 26
selection, 81–87
swept area, 18
used, 85–86, 112
See also household-size turbines; micro turbines;
mini turbines
solar modules. *See* photovoltaic modules
solar/wind systems, 43,
44–45, 46
Southwest Windpower
Air 303 (turbine), 2, 13,
32–34, 75
brake switch, 77
controls, 83
furling technology, 27
generators, 28
performance, 10, 30
rooftop mounting, 40–41
tower kits, 72
Windseeker (turbine), 27, 74
Starrs, Tom, 104
storage heaters, 52
strain relief wire net, 94
subsidies, 87
swept area, 7, 9–10
performance determina-
tion by, 18–20
of small turbines, 18
swept area estimation
method, 18–20

T
tail vanes, 26
telecommunications, 56, 57–59
Third World. *See* village electrification

towers
climbing, 78–79
height, 14–15, 71–75
lowering, 79
placement, 74, 75, 78
raising, 79, 89–90, 93, 95
safety, 77–78
selection, 82–83
types, 40, 41
See also guyed masts
Trace inverters, 45, 47, 65, 77
Tractel griphoists, 90, 92, 93,
99
treetop mounting, 73–75
Trojan batteries, 45, 47
tubular towers, 40, 41
turbines. *See* wind machines
Turek, Steve, 38, 86

U
U. S. Department of Agriculture (USDA), 19
United States, 43, 47, 51
incentives, 87
net metering, 67
potential, 103
wind power plants, 1, 61
University of Massachusetts,
50
urban wind, 75–76
utility interties, 34, 61–69
degree of self-use, 65–67
European, 43, 67–69
excess generation, 17
net metering, 63, 67, 104
power quality, 67
technology, 63–65

V
vehicles, electric, 57
Vergnet, S. A., 58
furling technology, 27

generators, 28, 63
tower kits, 93, 94–95
vertical furling, 27–28, 29
village electrification, 58
batteries, 45
demand, 104
power needs, 59

W
water pumping. *See* pumps
Westwind Turbines, 38, 94
Wilkins, Paul, 89
winches, 92
windchargers, 2–3, 6
wind-electric pumping, 48
wind energy associations, 12,
113–14
wind furnaces, 50
wind machines
configuration, 25–29
interconnected systems,
61–69
medium-sized, 1, 4
new designs, 38–40
noise level, 75
safety, 76–79
sizes, 13
solar systems with, 44–45
stand alone systems, 44,
48–59
See also small turbines
wind measurement devices,
8, 9, 12, 112–13
windmills, farm, 48, 49
wind power
air density, 7–8
swept area, 7, 9–10
wind resources, 11–12
wind speed, 7, 8–11, 12–15
wind power plants, 1, 61, 86
wind pumps. *See* pumps

wind speed
  distribution, 10–11
  height and, 12–15
  power and, 7, 8–9
Wind Turbine Industries, 27,
    28, 38, 64
wind turbines. *See* wind machines
  chines
work platforms, 78, 79
World Power Technologies,
    13, 35–36, 46, 75
  controls, 77, 83, 84
  furling technology, 27, 29,
    35–36
  power estimation, 23
  tower kits, 72, 84
  Whisper 600 (turbine), 2,
    59
  Whisper 1000 (turbine), 13,
    38, 47
  Whisper H1500 (turbine),
    38
wound-field alternators, 28
Wulf, Ed, 51

Y
yachts. *See* sailboats

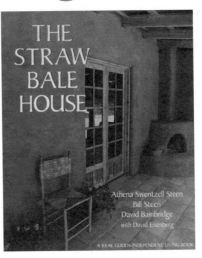